Garden Bird *Confidential*

hamlyn

Garden Bird *Confidential*

Discover the hidden world of garden birds

Dominic Couzens

An Hachette UK Company
www.hachette.co.uk

First published in Great Britain in 2010 by
Hamlyn, a division of Octopus Publishing Group Ltd
Endeavour House, 189 Shaftesbury Avenue, London, WC2H 8JY
www.octopusbooks.co.uk

ISBN 978-0-600-62052-5

A CIP catalogue record for this book is available from the British Library

Printed and bound in China

10 9 8 7 6 5 4 3 2 1

CONTENTS

INTRODUCTION

This is a handbook of garden birds with a difference. In addition to all the basic details that would be included in any such book – habitat, feeding, breeding, and so on – each entry includes a host of surprising revelations, many of which have only recently been unearthed. You will never again look at your garden birds in the same way. For example, through DNA analysis it is now possible to reveal the complex relationships that many birds form, and establish how they choose desirable members of the opposite sex. Such surprising revelations have turned our understanding of bird biology on its head; hence the 'confidential' information for each bird.

So what do your garden birds really get up to when you're not looking? This book has the answer: its purpose is to translate into easily digestible form the scientific research done into bird biology in the last 40 years. It covers garden birds throughout northern Europe, although inevitably any guide containing 60 species will end up with someone's favourite being left out. Hopefully, however, you will find most of Europe's regular garden birds within these pages. The description of each species covers the topics described below.

IDENTIFICATION

This section will help you recognize a bird by its plumage, and describes how to distinguish *male* and *female*, together with *juveniles* and, where appropriate, birds in *breeding* and *non-breeding plumage*. It includes the time of year in which you are likely to see the bird in the plumage described.

Juvenile plumage is the first plumage worn by a bird that has left the nest. It is usually moulted in late summer, by which time the youngster resembles the adult. However, in some cases a bird a few months old remains distinguishable from the adult in the winter months, in which case it is referred to as a *first winter*. A few garden birds take several years to mature and the blanket term *immature* is used for their various plumages.

SHAPE AND CHARACTER

Many birds are easier to identify by shape or behaviour than by plumage – classic examples being the treecreeper and the black redstart: small brown birds that, respectively, shimmy distinctively up trees in mouse-like style and habitually shiver their tails. In this section there are hints to enable you to identify birds easily.

VOICE

Bird sound recognition is hard to master, but invaluable for identification. Written descriptions cannot substitute for hearing the real thing, but this section gives some tips and mnemonics to help you identify them – ideally in conjunction with a CD of bird sounds.

There are two types of sounds, *calls* and *songs*. Calls are the everyday language of birds, usually just a couple of syllables repeated, and often given in flight. Songs are typically 'sentences', phrases or rambling sequences of sounds, more complex than calls. They tend to be given only in the breeding season and are associated with

◁ *You can sex this goldfinch as a male because it is feeding from teasel. Females have bills that are fractionally shorter, making them less able to reach seeds inside the flowerhead of this plant.*

◁ *Did you know that, in order to build their nests, long-tailed tits need a large feather supply, and often use corpses of birds.*

▽ *Jackdaws are the 'smug marrieds' of the bird world, forming an exceptionally close and faithful pair-bond.*

territory and mate attraction. The season when such songs are heard is mentioned in the text, but only where it is useful and only for resident species. Summer visitors sing as soon as they arrive in the spring.

HABITAT
All the birds described in this book occur in gardens, but which types of garden? The general habitat preferences of the various species are given here.

FOOD
This section describes the main dietary preferences of the bird in the wild, which are often seasonal. For garden feeding options, see the next section.

IN THE GARDEN
Bird husbandry is the theme here. How can you help a bird in the garden? What are the right foods and feeders to put out, and how might a species be encouraged to breed within your borders?

BREEDING
Whether or not a certain bird breeds in gardens, its breeding statistics are given in this section. A short description of the nest, its site and materials is generally included, along with the number of eggs, the incubation period and the time a bird takes before it leaves the nest (fledges). Sometimes there are notes on how the sexes divide breeding duties between them, what the main breeding season is and, where appropriate, the care of young after they have left the nest.

MIGRATION
A bird's appearance in the garden is often highly seasonal, and this section aims to give some idea of when birds appear where. Many, such as swifts, occur only in the summer months, while waxwings are deep-winter visitors to most of Europe. This section can only offer a rough guide – birds make a habit of breaking the rules about where and when they are supposed to occur.

ABUNDANCE
Abundance is a can of worms, because a bird that is common in one place may be rare only a short distance away. However, this section gives a general picture of how frequently certain species are found in gardens.

THE AMAZING WORLD OF BIRDS
In conclusion, I hope you discover here fascinating facts about your own garden regulars that will not only broaden your understanding of their lives, but also increase your respect for these amazing creatures. In many ways the book is a tribute to the scientists who, through their painstaking research, have made the discoveries that astonish and inform us. But to be honest, the birds are the stars: we are privileged to share our space with them.

Feral pigeon

Species: *Columba livia*
Family: Columbidae

IDENTIFICATION The urban pigeon (31–34 cm/ 12¼–13½ in) comes in all kinds of colours. The eyes are red. Most variations have striking pale grey under the wing and two black wing-bars on either side of a white rump, and almost all have an iridescent turquoise/purple patch on the neck.

The *male* and *female* look alike.

The *juvenile* (all year) has slightly browner plumage on grey birds, with less (or no) glossy colour on the neck.

SHAPE AND CHARACTER The pigeon looks plump on the ground, but often quite lean in flight, with long wings and tail. It is a very sociable species that lives in permanent flocks, often in city parks, coming readily to eat bread or grain. Extremely gregarious, it lives in permanent colonies that have their own 'patch'. It is often seen displaying on the ground, the males pursuing permanently disinterested females and bowing and ruffling their feathers. The pigeon has a brief aerial display, flying out and gliding on raised wings.

VOICE Birds displaying on the ground make a stammering sequence of coos; otherwise, the pigeon's coos have a moaning quality. The pigeon also makes various ecstatic coos at the nest.

HABITAT This is very much an urban bird, rarely seen far from buildings, and common on roofs.

FOOD It eats seeds and grain and, in cities, almost anything that is vegetarian.

IN THE GARDEN The pigeon is harder to discourage than encourage. It will take bread and grain and all manner of scraps from level bird tables and ground stations. Dovecotes encourage pigeons to breed.

BREEDING This is one of the few birds that may breed in any month of the year, during which time it may raise five or more broods almost continuously. It builds hardly any nest and the clutch is two eggs, which are incubated by both sexes for 17–19 days. The young leave the nest when they are 35–37 days old.

MIGRATION Pigeons are mainly sedentary, although racing pigeons can be taught to return to their home loft from hundreds, and even thousands, of kilometres away. Training and travelling flocks often pass over gardens.

ABUNDANCE These birds are common in gardens, and abundant in urban ones.

◁ *Feral pigeons are extremely sociable and live in stable flocks. Individuals in such flocks will know the rest of the flock members as well.*

▷ *The feral pigeon has the most variable plumage of any pigeon species, but the eyes are always red.*

FERAL PIGEON

Flight plans

Researchers from the University of Oxford, England, were surprised recently when they put miniature cameras on homing pigeons in order to see what visual cues they might use to navigate back home.

It has long been known that pigeons (and other birds) use landmarks when travelling. But even so, the scientists were astonished to find that, whenever possible, birds fly over … roads. True, we have roads to help us get from place to places, but we are land-based. What use could roads be to pigeons? Well, it turns out that they are simpler to use than other cues – especially major roads and motorways, which tend to run in straight lines. They are so convenient, it seems, that when faced with the choice between a direct route and navigating by means of a road, pigeons frequently choose the latter option. They even follow motorways as far as certain junctions and then turn off at the appropriate one. On major roads pigeons will turn off at appropriate roundabouts. Evidently, even homing pigeons keep up with the times.

◁ *Almost uniquely among birds, pigeons feed their young on a kind of 'milk' secreted by the crop.*

The mystery of baby pigeons

It's a common question asked by everybody: 'Why, when there are so many pigeons about, don't you ever see their babies?'

The answer lies in pigeons' development. When they are recognizably 'babies' – that is, smaller than their parents, and cute and downy – pigeon young are confined to the nest, where people don't see them. After a month or so there, they are so well grown that they are the same size as their parents and not especially different in plumage. Often, they just look a bit bedraggled. So even though they might be 'babies', in the sense of being juveniles and new to life, they don't look like them. Hence the 'mystery' of the missing progeny.

Driving me crazy

Everybody has seen the pleasing strutting and cooing of male pigeons as they display to females in city parks. Much less well known is a less chivalrous performance known as 'driving'. In this case, a male runs closely behind its mate, almost stepping on the female's toes as it moves forward. The display has its purpose – to prevent the female indulging in any extra-pair copulation. By walking closely on the female's tail, the male can ensure that it won't be able to sneak out of sight.

Pigeon talents

It is easy to underestimate pigeons, perhaps because they are familiar and apparently a bit gormless. But these birds have been intensively studied, and it has been shown that they have some extraordinary talents that you and I could only dream of possessing.

One such talent is the ability to detect extremely low-pitched sounds that are inaudible to the human ear. We can hear down to about 20 Hz, but pigeons' hearing ranges down to a fraction of this – to 0.05 Hz. In theory, this enables them to pick out subsonic rumbles, which radiate, for example, from the jet stream, from ocean waves and from the wind blowing against topographical features such as cliffs and buildings. It is possible that different parts of the world have a sort of 'sonic profile', and some scientists go so far as to say that pigeons could use these infra-sounds to give them an idea of where on the planet they are.

Another pigeon talent is a great sensitivity to barometric pressure, probably by means of the inner ear. So sensitive are they that the birds can, in theory, tell when they have gone up just 10 m (33 ft) in altitude. Quite apart from giving them the ability to know how high they are flying, this also presumably enables pigeons to predict when the weather is about to change.

Pigeon droppings

Birds don't excrete in the same way that we do. For a start, they combine urine and faeces into one package to produce a paste-like substance. This helps in water conservation, but it can make bird poo troublesome...

The average pigeon excretes 2.5 kg (5½ lb) of guano a year, and there are buildings in the world's urban centres that have collected up to 1 m (3⅓ ft) of deposits. Pigeon droppings carry diseases, including chlamydiosis (a respiratory disease), bird flu and even a form of meningitis. They are also corrosive. In 2007 a bridge collapsed in Minnesota, killing 13 people, and the steady degradation of the structure by chemicals in pigeon excreta has been suspected as a possible cause.

▽ *Pigeons are among the swiftest-flying of all birds.*

Wood pigeon

Species: *Columba palumbus*
Family: Columbidae

IDENTIFICATION Europe's largest pigeon (40–42 cm/ 15¾–16½ in) is easy to identify, with its combination of white neck patch, yellow eyes and prominent white crescents that cut across the wing (best seen in flight). Note also the pale edges to the primaries (the main flight feathers that form the wing-tip) when the bird is perched.

The *male* and *female* look alike.

The *juvenile* (mid-spring to early autumn) looks like the adult, but lacks the white or iridescent green on the neck.

SHAPE AND CHARACTER The wood pigeon is larger, longer-tailed and more deep-chested than other pigeons. On the ground it has a waddling walk and looks heavy. It often flies off with a clatter of wings, losing the odd feather. The aerial display is a common sight: the bird takes off with a clap of wings, rises to a height, stalls and then glides down – this differs from the display of the collared dove by being forward-moving, rather than straight up and down.

VOICE It has a distinctive, very throaty coo of five syllables in a distinct rhythm: quick-slow-slow-quick-quick. The phrase, which can be rendered as 'Oh dear, how boring', can be repeated several times in a row.

HABITAT At heart this is a shy woodland bird that feeds in fields. Nowadays the wood pigeon has transformed itself into a garden and urban regular, perching on chimneypots, overhead wires and roofs, and generally being common and easy to see.

FOOD The wood pigeon takes a variety of vegetable items, including seeds, leaves, buds, berries and roots. At times it is a serious agricultural pest. It is fond of ivy berries, acorns and clover leaves.

IN THE GARDEN Most people prefer to discourage wood pigeons in the garden, but they come to bird tables for bread, seeds and other scraps. Wood pigeons also eat beans and peas.

BREEDING The main breeding season is very late, between mid-summer and early autumn, although it has been recorded breeding in every month of the year. Both sexes build a flimsy-looking platform of twigs and incubate for 17 days; the young fly 33–34 days after hatching – sooner if they are disturbed.

MIGRATION Wood pigeons tend to be present all year, but this masks local and national movements. Large flocks are often seen in mid- to late autumn, when thousands of birds move out of Scandinavia and eastern Europe.

ABUNDANCE It is often exceedingly common.

◁ *Largest of the pigeons, the wood pigeon has a white patch on the side of its neck and a pale eye.*

WOOD PIGEON

▷ *The nest here is sturdy for a pigeon. Many structures are flimsy at best.*

Flock dynamics

Wood pigeons often form large flocks and a fierce hierarchy dictates where individuals can feed. Dominant birds feed in the centre, while subordinates – often young birds – have to feed around the periphery, where they are much more vulnerable to predators and are frequently picked off over time by a range of enemies. Furthermore, a place on the flock edge means spending more time looking for danger and less time actually eating. In such a predicament birds may quickly weaken and become even more vulnerable to predators. There isn't much altruism in a wood-pigeon flock.

Seasons and turns

The wood pigeon has an unusually late breeding season, with a peak of territorial activity in mid-summer. Males that are singing and apparently ready to breed in late spring often don't do so.

Recent research has shown that incubation shifts may contribute to this. When eggs are in the nest, the male incubates between 10.00 and 17.00 hours, with the female taking over for the remaining time. This gives the male nine hours and the female seven hours of feeding time a day. Earlier in the year, when this much time is not available, the incubation shifts are frequently interrupted, causing the adult to take time away from the nest and leave the eggs vulnerable to predation.

Pigeon milk

Pigeons are exceptional among birds for feeding their young on the equivalent of mammalian milk, a protein-rich secretion produced in the crop, which the young effectively 'drink' from the parents' throats. It is quite different from the fresh food brought to the young by most birds, or from food that is half-digested and then regurgitated. Among the world's birds, only pigeons, emperor penguins and greater flamingos feed their young on such milk.

The pigeon's milk is full of protein, fat and minerals, and is fed to the youngsters from day one. Both parents produce it, and do so for 16 days. Gradually the parents encourage the chicks to take some seeds and leaves from the crop as well as milk, and slowly 'wean' them onto an adult diet.

Collared dove

Species: *Streptopelia decaocto*
Family: Columbidae

◁ *Collared doves are extraordinarily productive birds, often attempting four or more broods in a season.*

IDENTIFICATION The collared dove (32 cm/12½ in) is smaller than the pigeons, with a long tail and pale, creamy plumage. The clearest field mark is the single black neck bar, which has subtle white edges.

The *male* and *female* are alike.

The *juvenile* (all year) lacks the black collar.

SHAPE AND CHARACTER The collared dove is a streamlined dove with a relatively long tail. It shares with the turtle dove (see pages 16–17) a characteristic 'flicking' action of the wings when flying, in contrast to the more powerful, steady flight action of the larger pigeons. Common and tame, it is often seen sitting atop roofs and aerials in gardens. It has a characteristic display flight, rising up steeply from a high perch with a few wing-claps, stalling and then floating down on spread wings and tail, describing a downward spiral.

VOICE This is often a dominant sound in suburban neighbourhoods. It makes a repetitive three-note coo, one short syllable followed by two long syllables, easily rendered as 'U-ni-ted'. It also makes an unusual sound upon alighting – a higher-pitched, more vibrant call, like the sound of a party trumpet.

HABITAT The collared dove is very much a suburban species that thrives in gardens; it needs dense trees for nesting. It is not common in truly 'wild' places.

FOOD It is mainly vegetarian, taking a variety of seeds and grains, and occasionally berries.

IN THE GARDEN It comes readily and with little encouragement to feeding trays and ground stations where seed is provided.

BREEDING The collared dove may lay its clutch of two eggs at almost any time of the year and carries on production – conveyor-belt style – throughout the summer, recording three to six broods. It builds a platform-nest of fine twigs and plant stems in a small tree. Both sexes incubate for 16–17 days and the young leave the nest 17–19 days after hatching. In common with other pigeons and doves, the adults feed the nestlings on crop milk.

MIGRATION This is a sedentary bird, making just a few local movements.

ABUNDANCE It is very common.

COLLARED DOVE

Coos and effect

The familiar three-note coo is not universal in collared doves; some males deliver two-note coos, especially at the end of a calling bout.

Recent research has shown that the number of two-note coos is related to the size of the male. The lighter, weaker males are most likely to offer the abbreviated version, which suggests either that they are not fit enough to coo properly or that they simply can't be bothered. Either way, selecting a good male should be a no-brainer for the females.

The invader

It might look comfortably settled in its suburban niche throughout Europe, but over much of Europe the collared dove is a comparatively recent invader. Originally a bird of India and the Middle East, it was virtually unknown in Europe until the beginning of the 20th century, when it started perhaps the most rapid and far-reaching expansion of range ever recorded for a bird. It had spread to the Balkans by 1928, to much of central Europe by 1957, to Britain by the mid-1950s, Italy by 1963 and northern Spain and southern Fenno-Scandia by 1977. Between 1930 and 1970 the collared dove is thought to have colonized 2.5 million sq km (965,000 sq miles) of land and to have jumped from zero pairs to 14 million pairs in 70 years.

Astonishingly, we still don't know exactly why the collared dove appeared from nowhere to become one of continental Europe's most familiar birds. We do know that it has a large brain, with a tendency to attempt novel forms of foraging and to wander as a juvenile bird – but none of this can quite explain its extraordinary population explosion in Europe.

The European invasion is only part of this bird's phenomenal success. After being introduced to North America, it is now rapidly spreading there.

Cowboy builders

Whatever the success of the collared dove, it is fair to say that it probably isn't down to this bird's nesting abilities. There are many records of youngsters actually slipping through the nest structure and down to their death below.

▷ *Shunning forests and other 'wild' habitats, the collared dove really is most at home in the suburban landscape.*

Turtle dove

Species: *Streptopelia turtur*
Family: Columbidae

IDENTIFICATION The turtle dove (26–28 cm/ 10¼–11 in) is smaller than the collared dove and is an exquisitely marked, slimline dove of agricultural areas. The beautiful tortoiseshell markings on its back are diagnostic (that is, identification certainties), as are the curious 'zebra-crossing' stripes that make up its neck patch. It has a black tail with a narrow white rim, which is conspicuous when the bird is landing. Its underwings are dark, contrasting with the pale belly. Its breast is stained a very light pink.

The *male* and *female* are alike.

The *juvenile* (mid-summer to early autumn) lacks the black collar at first, but this is gained by autumn, though it is smaller than the adult's. At first it is colourlessly scaly on the wings, but acquires a golden colour patchily.

SHAPE AND CHARACTER The turtle dove is more delicate than other pigeons and doves. In flight it is a streamlined dove with a characteristic 'flicking' action of the wings; it also has a habit of pitching intermittently to one side, then the other. It looks fast and manoeuvrable. It is not as sociable as other pigeons and doves, but is still usually seen in small groups. In common with the collared dove (see pages 14–15), it performs a pleasing 'up and down' display, flapping its wings hard to reach height and then sailing down.

VOICE It makes a unique, pleasing, soporific purring, with a few purrs uttered in a short series.

HABITAT It inhabits lowland farmland, scrub and hedgerows and is often seen perching on overhead wires. t is common in many rural villages.

FOOD The turtle dove eats the seeds of various arable-ground and waste-ground weeds, together with some grain. It feeds on the ground – not normally from bushes or trees.

IN THE GARDEN Where this bird is common, it may come to ground stations for grain.

BREEDING This bird builds a platform of thin twigs, lined with finer materials, in a shrub, bush or small tree. It lays the usual pigeon clutch of two white eggs, both sexes incubating them for 13–16 days. The young leave the nest after about 20 days. It may attempt two or even three broods in a season.

MIGRATION In complete contrast to other pigeons, he turtle dove is exclusively a summer visitor, arriving in mid-spring and leaving by early autumn. It winters south of the Sahara.

ABUNDANCE Locally this bird is common.

◁ The stunning scalloped back and zebra stripes on the side of the neck make the turtle dove unmistakable. The name 'Turtle' comes from the 'tur, tur' song.

▷ In common with other pigeons and doves, the turtle dove feeds mainly on the ground.

TURTLE DOVE

Weed lovers

Studies in Britain have shown that the turtle dove has surprisingly specific tastes. Although it is known to eat grasses of various kinds, plus grain and chickweed capsules, its distribution is strikingly correlated to a somewhat obscure plant that is found in weedy fields and waste ground – the fumitory *Fumaria officinalis*. Very few European birds apart from finches have such strong links to a particular herb, although rather more depend on specific tree species.

Brood reduction

The turtle dove is in sharp decline over many parts of Europe and some recent studies suggest this has something to do with a change in its breeding seasons. A count of turtle doves on migration between 1963 and 2000 showed that, on average, these birds depart eight days earlier than they used to. Allowing for preparation time, this takes about 12 days off their breeding season and this could mean the difference between bringing up two broods or just one, particularly in northern Europe. As yet, nobody seems to have worked out why this change has occurred, but it could be that the turtle dove is now out of synch with the time of year when its food is most abundantly available. It should be noted that very large numbers are shot each year in parts of central and southern Europe, which has not helped its fortunes.

Turtles basking

For a bird that breeds in temperate climates, the turtle dove seems to have a surprising tolerance to heat. On its wintering grounds in Africa it is known for its habit of feeding under the glare of the midday sun without any ill effects. At one point it was observed foraging in an air temperature of 45°C (113°F).

Cuckoo

Species: *Cuculus canorus*
Family: Cuculidae

IDENTIFICATION This is a distinctive slim, grey, long-tailed bird (32–34 cm/12½–13½ in, the size of a pigeon) with a hawk-like appearance, but it is difficult to see. It has a small head, with a short curved bill and a yellow eye-ring. It has small yellow legs. Its plumage is slate-grey on all of the upper parts and breast, and white below with close, dark barring. Its underwings and undertail are also barred.

The *male* is white with dark bars on the upper breast.
The *female* has a tinge of buff on the upper breast.
The *female variant* is rusty-brown above, with barring.
The *juvenile* (early summer to early autumn) is barred both on the underside and on the upper parts, which are dark brown; the head is also barred, except for a white patch on the nape.

SHAPE AND CHARACTER The cuckoo's small head and long tail give it a distinctive shape. When perched, it often droops its wings and cocks its tail in the air; even so, it perches horizontally. Its flight is highly distinctive, with rapid wing-beats that are very shallow, the wings never seeming to rise above horizontal. It often flies low and far. The cuckoo is not a sociable bird. It often keeps well in cover, but will also perch prominently on wires and treetops.

VOICE This hardly needs any introduction: the male's song is a loud, far-carrying 'cuck-oo'. It makes variants of this call and a throaty chuckle. The female utters a loud ringing trill and sounds extremely excited.

HABITAT The cuckoo inhabits lowland farmland, scrub, marshes, woods – almost any habitat, including gardens. It is often seen perching on overhead wires.

FOOD It tends to eat caterpillars more than anything else, including the hairy types that are avoided by most other birds. It also eats beetles and ants, and will take a few eggs from other birds.

IN THE GARDEN This is a casual garden visitor.

BREEDING As everyone knows, the cuckoo delegates the care of its young to unwitting foster parents, namely small

△ *Most female cuckoos are grey and resemble males, but a small minority sport a rusty-brown plumage, like this bird.*

▷ *Cuckoos eat very large numbers of caterpillars. They easily manage the noxious and hairy ones that other birds avoid.*

insectivorous birds. Any given female may visit 50 nests and lay 20-plus eggs, one per nest. The female watches the nests for some time before striking. It lays the egg in the afternoon, and also usually removes an egg of the host bird. Once hatched, the nestling removes the rest of the host clutch (eggs or nestlings) by heaving them over the side of the nest with the small of its back. Having monopolized the parents' provisioning, it remains in the nest for about 19 days. Nearby parents of various species may bring it food, enticed by its loud begging.

MIGRATION A summer visitor to Europe, the cuckoo winters in tropical Africa. The adults arrive in Europe in the spring, and depart extremely early, certainly by late summer. The juveniles have gone by early autumn.

ABUNDANCE Locally the cuckoo is common.

△ *The reed warbler is one of the most frequent hosts of the cuckoo, especially in Britain.*

Female host preference

Although the cuckoo has been recorded parasitizing more than 100 species of birds in Europe, this figure masks the fact that cuckoos are usually fussy about which birds they burden with their eggs. For example, in each area of Europe local cuckoos have only a few regular hosts: in the British Isles it is reed warblers, meadow pipits, dunnocks and robins, while in central Europe garden warblers and pied wagtails are important, and in northern Europe redstarts and bramblings are the most common hosts.Furthermore, individual female cuckoos have certain specific preferences: namely, the species in which it was itself raised. So female cuckoos born in reed warbler nests grow up to lay their eggs in reed warbler nests. Only if the number of local nests is exhausted, or if it is a matter of specific convenience, will a female look towards alternative species.

Males, it is thought, have no such preferences. They form no strong pair-bond with any female, most of them being promiscuous. It is highly unlikely that they selectively copulate with reed warbler cuckoos or any others. The preference is carried along the female line.

Finding nests

Finding nests to parasitize isn't easy, even if you're a cuckoo. It usually requires a great deal of patient sitting around and observing before a cuckoo can locate a suitable nest and test when it will be ready to take eggs.

With some species, such as meadow pipits, the host nests can be so dispersed and well hidden that different tactics are called for. In this case, the cuckoo takes advantage of its unpopularity. It is quite routine, in the course of its day, for the cuckoo to be mobbed by smaller birds and potential host species, which call loudly, fly at it and attempt to drive it away – it's all part and parcel of the parasite's lifestyle. But the intensity of the mobbing can be a give-away: if the mobbing birds are moderately anxious, the cuckoo's threat can be gauged to be low; if they are extremely alarmed, the cuckoo can assume that the threat level is high and the nest is in the near vicinity, and it will look for the nest with particular care. Thus parasite and host play out a bizarre version of the 'Hotter, colder' game that we sometimes play as children.

△ *Cuckoos have a very distinctive manner of flight, the wings never seeming to rise above the horizontal plane.*

Home alone

Having contracted out the care of its young to a series of foster parents, the cuckoo's breeding season comes to an end as soon as its eggs are laid, or, in the case of the male, as soon as the females are no longer soliciting for copulation. Cuckoos have no need to hang around and find out the fate of their progeny. The youngsters have an internal clock and compass that will help them make their own way under their own steam to the wintering areas.

Thus, their task completed, some adult cuckoos migrate south as soon as they can. Amazingly, this means that some reach their winter quarters even before their offspring leave the foster nest.

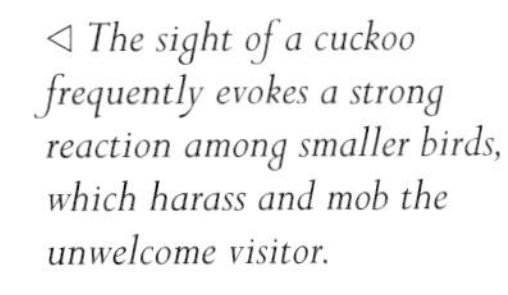

◁ *The sight of a cuckoo frequently evokes a strong reaction among smaller birds, which harass and mob the unwelcome visitor.*

Barn owl

Species: *Tyto alba*
Family: Tytonidae

IDENTIFICATION The barn owl (33–35 cm/13–14 in) is smaller than the tawny owl. A unique, ghostly white owl of the countryside, it is usually seen in the evening, flying low over fields. It has an unusual white, heart-shaped face, with black eyes that are small for an owl. Its plumage is pure-white beneath (but see the Variation below), its upper parts orange-buff and grey, peppered with small spots.

The *male* is usually greyer above than the *female*, which has a few spots on its flanks.

The *juvenile* (early summer to early autumn) may have some remnants of white down in the plumage.

Variation Birds in central, northern and eastern Europe have yellow-orange underparts, lightly freckled, and are darker grey above.

SHAPE AND CHARACTER The barn owl has a large head, a unique heart-shaped facial disc, small eyes and very long legs, which are often dangled in flight, combining to make it unique. It flies on fairly long wings, usually hunting low over fields or marshes and progressing with an unsteady, wavering flight, punctuated by regular hovering, sharp turns and two-footed dives. It is essentially nocturnal, but can be seen at twilight, particularly in winter. It is a solitary bird.

VOICE It makes a blood-curdling screech. The young make various hisses and gargles.

HABITAT This bird inhabits lowland farmland with hedges and ditches. It is common in many rural villages.

FOOD It is chiefly a predator of small mammals, including mice, voles and shrews, but also takes a few birds and amphibians. It hunts low down, following transects over fields and field edges, and locates most of its prey by sound.

IN THE GARDEN In large rural gardens, barn owl populations can be helped by the provision of nest-boxes.

BREEDING The barn owl selects a cavity in a building such as a barn or tree, or on a cliff or among rocks for nesting. It makes no nest as such, simply laying four to seven white eggs straight onto the base of the cavity. The eggs are incubated for 27–34 days by the female while the male brings in provisions. The young fly at about 60 days of age and may be dependent for two months or more. Nevertheless, some pairs are double-brooded.

MIGRATION This bird is not a migrant, but recently fledged young can disperse over considerable distances.

ABUNDANCE Locally the barn owl is common.

◁ *The barn owl has smaller eyes than other owls, suggesting that hearing plays a major part in its hunting lifestyle.*

▷ *One of the more nocturnal owls, the barn owl is only occasionally seen by day, and then usually at dawn and dusk.*

BARN OWL

Amazing hearing

Owls are famous for their eyes and night vision, but in fact this talent is exaggerated. They don't see all that much better in the dark than humans do, although they get a lot more practice. What really sets an owl apart from a human being is its hearing, which is many times more acute than ours. Few animals hear more acutely than the barn owl. Its health and life depend on it detecting minute rustles and squeaks and then diving upon the creatures that make them.

Several features make a barn owl's hearing special. One is that their ears are large. Another is that ridges on the face deflect sounds onto the ears (on the outer edge of the facial disc), amplifying the signal. But the weirdest adaptation is that the left ear is slightly higher than the right, which would make a human being look decidedly unbalanced. However, this asymmetry gives the barn owl's hearing a three-dimensional sensitivity, which can be understood if we look at how we detect the direction of sounds. For us to be able to tell where a sound is coming from, we rely on the minute time difference that sound waves take to reach one ear relative to the other – if a sound is coming from our right, for example, it reaches the right ear sooner than the left, and the difference is enough to inform our perception. Not only does the barn owl share this ability, but its asymmetrical ear openings also enable it to perceive differentials in the horizontal plane – any sound coming from above will hit the left ear before the right, and vice versa. This three-dimensional sense of hearing enables the barn owl to hunt in total darkness.

Spotting talent

Male barn owls are turned on by spots. In this species, males are not strongly marked on the underside, but females are, and the number and size of their spots vary between individuals. A study in Switzerland showed that it isn't the number of spots that matters – a female is not more attractive for having more spots or fewer. It is all a matter of personal choice. Certain males are attracted to a certain degree of spottiness. Furthermore, it seems that preferences are passed down the male line, with fathers and their sons being attracted to the same kind of female.

◁ *Astonishingly, the barn owl is capable of catching fast-moving prey in total darkness.*

△ *The strange, ghostly white plumage of the barn owl is certainly distinctive. Some think that it actually glows in the dark in some unknown way (bioluminescence).*

Disproportionate production

In a US study scientists found that breeding success in the barn owl is a highly individual affair. Some birds, you might say, are 'good at' breeding, whereas others are little short of hopeless – and that result comes from measuring a lifetime's reproduction, not accounting for inexperience or a season of bad luck. It seems that just a few pairs produce a high proportion of a population's chicks.

The scientists found that, in their study population, a mere 8 per cent of egg-laying females were responsible for 25 per cent of all eggs laid, and that 22 per cent of egg-laying females produced no fledglings at all. Amazingly, only just over 10 per cent produced any descendants – birds that went on to have young themselves.

Tawny owl

Species: *Strix aluco*
Family: Strigidae

◁ *Owls take readily to nest-boxes, because natural holes in trees large enough for them tend to be at a premium.*

IDENTIFICATION The tawny owl (38 cm/15 in) is about the length of a wood pigeon, but looks bigger. It is a warm-brown, richly coloured and patterned owl with black eyes and a rounded head, and narrow white eyebrows. It is often the most familiar owl in gardens.

The *male* and *female* look alike.

The *juvenile* (late spring to late summer) is rounded and fluffy, with greyish barred down, and has more prominent black eyes.

SHAPE AND CHARACTER The very large head is neatly rounded. This is quite a compact owl, with a short tail and very rounded wings. It is truly nocturnal, roosts secretly by day and is difficult to see. It flies with rather stiff wing-beats, without a sound, on a level (not undulating) course.

VOICE Many tawny owl calls are familiar from movie soundtracks. The male delivers an atmospheric series of hoots with a mournful or eerie ring. Usually there is an introductory hoot, a short pause and then a series of quavering hoots. In addition, both sexes give a sharp 'ke-WICK!' contact call, and the young deliver a squeak sounding a little like water rapidly going down a sink.

HABITAT This is a woodland bird that has adapted to gardens. It needs large, mature trees for roosting and nesting. It also favours lawns, driveways and roads that make it easy to catch its prey. It will hunt by the light of street lamps.

FOOD The tawny owl is a predator, taking a variety of living creatures including small mammals (mice, voles and shrews), birds, insects and worms (the last often from lawns). It hunts by watching from a perch and then pouncing down on whatever it detects moving. It will also catch prey in flight. It coughs up the hard remains of food – bones, skin, feathers and chitin (the hard parts of insects) – as pellets.

IN THE GARDEN It takes readily to nest-boxes, placed high up, but requires a large (for example, 76 cm/30 in tall, 20 cm/8 in square) open-fronted nest-box. Do not disturb the aggressive breeding birds.

BREEDING The tawny owl begins breeding early in the year, with many pairs having eggs in early spring. It just lays its eggs onto the bottom of a hole – usually in a large, mature tree high up. The clutch of two to three white eggs is incubated by the female, which is fed regularly by the male, for 28–30 days. The young fledge 35–39 days after hatching, but are looked after by the adults for months afterwards.

MIGRATION The tawny owl is a very sedentary bird; pairs remain on the same territory all their lives. The young disperse locally.

ABUNDANCE Tawny owls are fairly common.

TAWNY OWL

Too wet to 'woo'

For a long time birdwatchers have noticed that tawny owls are much quieter in the rain. Their quavering hoots are heard less often on damp nights than on dry nights, even at the height of the hooting season in the autumn. A recent study has shown why: the rain literally dampens the transmission of the sound. Apparently an owl's hoot travels 69 times further when it is dry than when it is wet, owing to the difference in atmospheric conditions.

Owls are not overly fond of rain. Their main hunting sense – sound – is obviously rendered less effective by the increase in background noise, together with the wind that often accompanies a shower. There is some compensation, though. On wet nights worms come more readily to the surface of lawns. They are nutritious and easy to catch, and tawny owls sometimes eat them in large quantities, bumbling around the lawn like giant blackbirds.

▽ *Tawny owls have soft plumage that helps them fly without any wing noise.*

Silent flight

Owls are well known for their silent progress through the air. They have an ability to appear suddenly without warning – a talent that they put to good use when hunting.

The wings of owls are specially adapted for noiselessness. They are invariably large for the size of bird, making flying easy. They also have several unusual plumage features. First, the feathers are all soft on the top surface, fitted with a downy layer, and this cuts down the noise they would otherwise make as they rub together. Second, the leading edge of the wing has a fringe like the very short teeth of a comb and this apparently helps the air to move smoothly above and below the wing. Third, the trailing edge of the wing is soft and dampens any turbulence.

The silence of the owl's flight is not only important for enabling a stealthy approach; it also helps the hunter not to drown out external noises by the sound of its own wings. Imagine trying to hear a mouse while walking along wearing noisy waterproof trousers and you'll appreciate the point.

Swift

Species: *Apus apus*
Family: Apodidae

IDENTIFICATION The swift (17–18 cm/6¾–7 in) is always seen flying in the air and doesn't perch on wires or roofs. Its plumage is mainly dark brown, but with a whitish throat. It has a wingspan of 40 cm (15¾ in).

The *male* and *female* look alike.

The *juvenile* (mid- to late summer) is similar, but has a larger white throat patch and narrow white fringes to its wing-coverts.

SHAPE AND CHARACTER The swift's sickle or anchor shape is distinctive: a small head, sharply pointed and swept-back wings, forked tail (the fork is sometimes closed to a point). It tends to fly with long sweeps, often very high, with easy accelerations and changes of direction. It is a sociable bird, usually seen in parties. It cannot perch on wires or trees; it flies straight into its crevice nest site. Otherwise it is always seen in the air. It often flies high into the sky at dusk, its screams fading as it rises higher and higher.

VOICE It has a distinctive, slightly husky squeal.

HABITAT This bird uses the air space above gardens, often high up, but lower in poor weather. It often flies around the rooftops screaming. It may fly over any garden, but requires tall buildings, such as church towers, on which to nest.

FOOD It feeds on small flying insects, such as aphids and ants, as well as web-borne spiders floating in the air. Despite appearances, it does not simply fly with its mouth open hoping to catch insects, but targets each one individually. When feeding young, it collects up to 1,000 items at a time that are all mashed together into a ball in the throat.

IN THE GARDEN If you have a suitable eave, specialist swift-boxes will be of great benefit to these birds. Keep the entrance blocked until the birds arrive in the spring.

BREEDING Late arrival ensures a late start to breeding, usually in late spring. The swift is normally a colonial breeder and build nests on ledges or in holes, usually on older buildings. There are two to three eggs in the single clutch, incubated by both sexes for 19–25 days. The fledging period of the young is even more variable, at 37–56 days, depending on the weather.

MIGRATION A summer migrant, arriving in late spring and leaving in late summer, the swift is present in northern Europe for only 16 weeks, spending the rest of the year in tropical Africa south of the Equator. During spells of inclement weather it will undertake large movements even during breeding.

ABUNDANCE This bird is very common, but declining.

◁ *In ancient times people suspected that swifts didn't actually have any legs. The family name, Apodidae, means 'no feet'.*

▷ *Swifts cannot perch like another birds. All four of their toes are oriented forward, so all they can do is cling.*

SWIFT

Escape movements

Swifts are fair-weather birds, depending on the presence of small insects up in the air to nourish them – they cannot eat anything else. This means that bad weather is a major problem. Low temperatures and strong winds render insects unavailable up in the sky, forcing swifts to seek food in sheltered locations, particularly low over water, where insects will still be hatching.

However, an alternative is simply to evacuate an area, and this is what swifts sometimes do, even during the breeding season. Radar studies have shown that they will fly towards an oncoming depression, at first flying directly into the wind and then clockwise around it, so that they reach calmer weather sooner than calm weather reaches them.

At times these escape movements can take a swift on a round-trip of 2,000 km (1,240 miles) – the equivalent of a weekend break for us, you might say. Flocks of up to 50,000 birds may take part.

Longevity

Swifts are exceptionally long-lived for their size. An equivalent land bird, such as a pied wagtail, has a lifespan of two years once it has become an adult (maximum recorded age: 11 years). The swift, however, has an average life expectancy of nine years, and a maximum recorded age of 21 years, showing that its aerial lifestyle is a good deal safer than living on the ground.

△ *Wet and windy weather is a problem for swifts; even in mid-summer they may evacuate away from inclement conditions.*

△ *The long, narrow wings of the swift give it exceptional aerial manoeuvrability.*

▽ *The swift has a relatively tiny bill, but a very large mouth.*

Sleeping

Swifts always seem to be flying, and it is rare to see them land anywhere except at the nest site, even at dusk. This has long led to speculation that they must actually sleep on the wing. Observations from pilots have confirmed that birds often rise to great heights on summer nights in the short hours of darkness and appear to enter some kind of sleep.

Recently it has been shown that swifts exhibit a similar mix of wakefulness and sleep to that shown by dolphins. Half of the brain sleeps at a time, while the other half is alert. In this way the bird can rest without losing all its faculties – including the power of flight.

Brood reduction

Feeding young on flying insects is a perilous business in the often-fickle weather conditions of northern Europe. Frequently there is not enough food around on a given day, and young swifts are specially adapted to go without a meal for long periods – sometimes days at a time. They can survive on low rations for a month or more, and lose half their body weight.

But eventually something has to give, especially in larger broods of three or four. Trying to feed so many mouths can seriously compromise the adults' own health, and on such occasions they have to abandon the brood and feed themselves for a while. This often results in the death of one or more chicks. The upside is that the surviving brood members have a better chance of survival.

Green woodpecker

Species: ***Picus viridis***
Family: Picidae

IDENTIFICATION This is a distinctive large yellow-green bird (31–33 cm/12¼–13 in, the size of a pigeon). It has a bright-yellow rump that is easy to see on its low, strongly undulating flight. It has a long, dagger-shaped bill, a red crown and a white staring eye within a black mask. It spends much time on the ground.

The *male* has a red centre to the 'moustache' below the eye.

The *female* has an entirely black 'moustache'.

The *juvenile* (late spring to late summer) is a spotted version of the adults, with black spots on pale underparts and white spots on green upper parts. The moustache of the male is entirely red, that of the young female black.

SHAPE AND CHARACTER The green woodpecker has a characteristic woodpecker shape, with a long, straight bill, a large head and a stiff, short tail. Its flight is equally distinctive, with intermittent bursts of two to three quick wing-beats ensuring that its flight is heavily undulating. This bird spends a great deal of time on the ground, where it can feed quietly for many minutes on end. It also clings to the trunks and branches of trees in a typical upright woodpecker manner.

VOICE It makes a loud, panicky alarm call when flushed, roughly 'kyu-kyu-kyu'. In the breeding season it makes a gentler, more musical laughing series of similar notes, which drop in pitch.

HABITAT This is a woodland-edge bird, and is found in rural gardens with lawns, in areas with mature trees.

FOOD It specializes on ants, standing over anthills in the soil, digging with its bill and lapping up the insects with its sticky tongue. It also takes ants in the trees. It sometimes eats other insects, such as beetles and flies.

IN THE GARDEN If present, it uses lawns for feeding, and will occasionally take mealworms, fat or fruit laid out on the ground. It may sparingly use an enclosed nest-box with a depth of 38 cm (15 in) and a hole of 6.3 cm (2½ in), filled with chippings.

BREEDING Both sexes make a hole in a tree for nesting. Its entrance is often very low, just 1 m (3⅓ ft) above ground. Five to seven eggs are laid straight onto the bottom. The eggs hatch after 19–20 days, with both sexes incubating them. The young fledge after a further 21–24 days.

MIGRATION The green woodpecker does not migrate, but the young birds disperse in mid-summer.

ABUNDANCE This bird is fairly common.

◁ *A female green woodpecker feeds a chick. When well grown, the nestlings come right up to the hole entrance to beg.*

GREEN WOODPECKER

CONFIDENTIAL

△ *Green woodpeckers spend most of their time on the ground feeding on ants. This is a juvenile.*

Ants on the deck

The degree to which this bird relies on a single food source – ants – is remarkable. These insects make up at least 90 per cent of the diet, and on occasion nearly 100 per cent. Almost all the insects are obtained by ground foraging, either by attacking the obvious nests of species such as wood ants, or by hopping over the soil and finding the shallow ground nests of soil-living species. At least 21 species of ants have been recorded being taken in Europe.

In some ways the green woodpecker could not be anything but an expert in ant-guzzling: it has one of the longest tongues of any European bird, stretching a full 10 cm (4 in) beyond the tip of the bill – when retracted, it has to be fitted into a special tube that goes around the back of the skull and on to the crown. The tip of the tongue is flat and wide, and is capable of being moved independently, so that it can reach well into cracks and fissures in the soil. Large salivary glands ensure that it is always sticky, so that the insects adhere easily to the tongue.

Treasure hunting

Armed with such perfect equipment for ant-scooping, all the green woodpecker requires in order to be the perfect predator is the ability to find the anthills it needs. Here, too, it is more than capable of finding a supply throughout the year. Ants are active all year, and the green woodpecker has been seen digging through 1 m (3⅓ ft) of snow to get at an anthill.

The secret is that it remembers where the nests are, probably using landmarks. These birds are territorial and quickly work out where all the best hunting sites are; over the course of several years they become experts at knowing where to look for their favourite food.

Great spotted woodpecker

Species: ***Dendrocopos major***
Family: Picidae

IDENTIFICATION This boldly coloured black-and-white bird (22 cm/8¾ in) has a bright-red undertail and is usually seen in trees or coming to hanging feeders. It is basically black above, but has a large white oval 'paint-blob' on its shoulder, barred black-and-white wings and tail-sides, and white underparts without streaks.

The *male* has a red patch on the nape.

The *female* has a black nape.

The *juvenile* (early to late summer) has a red crown (larger and brighter in young males) and a pale, pink-stained undertail.

SHAPE AND CHARACTER This bird has the distinctive woodpecker shape, with a long, straight bill, large head and stiff, short tail. It typically clings upright to tree trunks and branches, moving upwards (or downwards) with two-footed hops, hugging the bark. It frequently hacks into the wood. Its flight is distinctive – heavily undulating, with periodic bursts of quick wing-beats.

VOICE Its most common call is a far-carrying 'chick', sometimes uttered many times in succession when the bird is alarmed. It also makes a drumming sound in spring, by knocking a suitably sonorous piece of wood up to 40 times a second. The drum is fast, with an 'attack' at the beginning, then a slight fade. It is very far-carrying and characteristic.

HABITAT This is a woodland bird found in gardens in less built-up areas. It tends to feed towards the tops of trees, and particularly on dead wood, at least in winter. In summer it will glean insects from leaves in the canopy.

FOOD It takes a wide variety of animal and vegetable food. Its most famous feeding method is to hack holes in rotting wood to obtain insects and grubs, especially wood-boring beetle larvae. It takes many caterpillars in the spring, and in autumn and winter mainly seeds and nuts. It often has favoured 'anvils' up in the branches – spots where the wood is hard and the woodpecker can lodge seeds in order to hack them open.

IN THE GARDEN The great spotted woodpecker readily comes to hanging feeders for nuts, suet, fat and seeds. It will sometimes take to an enclosed nest-box with a depth of 30 cm (12 in), a floor about 13 cm (5 in) square and a hole 5 cm (2 in) in diameter; fill this with wood or polystyrene chippings. If woodpeckers are attacking your tit-boxes (see pages 114–117), use a 'woodcrete' box (a material that combines the strength of concrete with the appearance of wood), or fit an aluminium cover over the hole.

BREEDING This bird famously makes its own hole by excavating through tree bark to make a cavity. It lays a clutch of four to six eggs in late spring, incubated mainly by the female for 14–16 days. The young leave the nest 20–24 days after hatching and the brood is then sometimes split between the two parents, with each taking complete responsibility for half the youngsters.

MIGRATION It is a sedentary bird, although occasionally birds from the north of Europe will move south in large numbers if the cone crop fails.

ABUNDANCE The great spotted woodpecker is a common bird.

◁ *Woodpeckers are very resourceful and agile feeders, as this photograph attests.*

▷ *The stiff tail feathers of a woodpecker keep it in position when it clings to a tree.*

GREAT SPOTTED WOODPECKER

CONFIDENTIAL

Bird-of-prey?

Great spotted woodpeckers often augment their chicks' diet of insects with a different kind of meat in the spring. The woodpecker's bill is an ideal tool for breaking into the tree-holes (or nest-boxes) of small woodland birds such as tits, whose young make a nutritious meal, both for the adults and their offspring. Of course, being experts in bark foraging, the woodpeckers can easily tell when eggs have hatched and nestlings are available. In some areas, these woodpeckers are major predators of youngsters and can have an effect on the prey species' population. Crested tits (see pages 114–117) are particularly vulnerable in some places, and blue and great tits (see pages 100–107) in others. Some woodpeckers have also learned to break into the 'hanging basket' nests of European penduline tits and feast themselves on these young, too.

Tapping in

You might not expect birds such as woodpeckers to be particularly intelligent, but laboratory experiments have shown them to be good learners. Captive birds, for example, have been able to learn simple codes to ask for certain foodstuffs. In one particular laboratory, the code was one tap for a pistachio nut, two for a cricket and three for a mealworm. The birds could also predict when a pistachio nut was about to be on offer, by recognizing a picture of one.

▽ *Bursts of wing-beats alternating with keeping the wings closed means that woodpeckers fly in undulating fashion.*

Drumming

It's easy to assume that the loud and very fast beating sound heard from woodpeckers in the spring is made by the act of making holes in the wood – after all, the noise does resemble that of a pneumatic drill. However, it has quite a different function.

The term 'drumming' is certainly apt. Think of a percussionist playing the drums in an orchestra; the beating is not intended to make a hole in the drum, but rather to create a pleasing noise. For woodpeckers the situation is exactly the same. The sounds they make are intended as signals, and the fact that they are confined to the early breeding season suggests they are the equivalent of a songbird's 'singing'.

It is thought that each individual bird has its own recognizable drum, and that the sexes make slightly different sounds too. Undoubtedly some birds are truly original: instead of hitting wood, they have been known to tap on drainpipes, glass milk bottles and the metal sides of machinery.

The birds do make sounds when excavating holes too, but the rhythm is slow, more like the sound of beating with a pickaxe. Whatever they hit, the target is chosen for the pleasing sound that it makes.

A sharp tongue

Although the great spotted woodpecker is most famous for using its bill as a specialized feeding instrument, its tongue is no less remarkable a tool. It is extremely long, able to extend no less than 4 cm (1½ in) beyond the bill tip and thus deep into cracks and fissures where insects are hiding. The tip of the tongue is also capable of independent movement – almost analogous to a human finger – and is sharp, used to impale soft-bodied insects. Bristles near the bill tip also help to hold on to reluctant prey.

◁ A woodpecker drums on a piece of bark. It strikes the wood up to 18 times a second.

△ Great spotted woodpeckers regularly raid the nesting holes of small birds in the spring.

Lesser spotted woodpecker

Species: *Dendrocopos minor*
Family: Picidae

IDENTIFICATION This is a woodpecker in miniature (14–15 cm/5½–6 in, slightly larger than a sparrow), always looking much smaller than the others. It is identified by a combination of its size, a buff-tinged breast with thin streaks and, in particular, the 'ladder' pattern on its back, which is barred black-and-white, without any large blobs. There is no red on the vent (lower belly).

The *male* is distinguished by its crimson cap.

The *female* lacks the colour on its cap: its forehead is white, giving way to black on the back of the crown and on the nape.

The *juvenile* (early to late summer) is similar to the adults, but has browner underparts with a few more streaks.

SHAPE AND CHARACTER This bird has the typical woodpecker shape (large head, stiff tail, upright stance against vertical trunks and branches), but is just a lot smaller. It also has a smaller, less powerful bill than the great spotted. It is often a bird of the upper branches, foraging right out to the ends of the branches. It flies in typical undulating woodpecker style, but usually lands near the top of a tree. It can be extraordinarily elusive.

VOICE Its gentle 'chip' sound is barely distinguishable from the great spotted woodpecker's call, but the male also has a loud, quite explosive 'pee-pee-pee...' call – all on one pitch, sometimes slowing at the end. It drums in the same way as the great spotted, knocking its bill on wood to make a 'song', but the drum (mid-winter to late spring) lasts a little longer and doesn't fade.

HABITAT It inhabits open broad-leaved woodland, frequently with trees such as birch and alder, as well as orchards, parks, gardens and hedgerows.

FOOD It is less catholic than the great spotted, subsisting almost entirely on insects and their larvae. It pecks at rotten wood in winter, and in summer feeds mainly by gleaning from twigs and leaves. It will sometimes sally out to catch insects in flight, but doesn't forage on the ground.

IN THE GARDEN This bird is an infrequent visitor to feeding stations, for nuts and seeds and sometimes fruit. It occasionally uses enclosed nest-boxes (see page 34).

BREEDING It excavates its own nest-hole (both sexes), often higher up than other woodpecker holes and on thinner branches; it has no lining. It lays four to six white eggs, which are incubated by both sexes (the male sits at night) for 11–12 days. The young leave the nest 18–20 days later, but are fed for some time afterwards outside. It has one brood a year.

MIGRATION This bird is not a migrant, but occasionally moves out of northern areas in the autumn.

ABUNDANCE Locally it is fairly common.

◁ *The streaks on the lesser spotted woodpecker's breast, together with the lack of any red on the belly, distinguish it from the much commoner great spotted woodpecker.*

LESSER SPOTTED WOODPECKER

CONFIDENTIAL

Mixed flocks

As if to reaffirm its credentials as a small bird, the lesser spotted woodpecker is a regular member of the roaming parties of birds that are such a characteristic feature of the woodland in autumn and winter. Joining up with long-tailed, blue and great tits, goldcrests, nuthatches and treecreepers, it tours woodlands under the protective veil of a great deal of collective watchfulness. In fact, it is not unknown for great spotted woodpeckers to do the same.

The benefit of such an association is that, while each species more or less has the same predators – sparrowhawks, tawny owls and the like – they forage in their own micro-habitats and do not step on the toes of other birds. Thus while great tits in a mixed-species flock might feed on the ground, the constituent lesser spotted woodpecker will be in the thin twigs up in the treetops.

△ *Lesser spotted woodpeckers tend to inhabit the smaller twigs and branches high in the tree canopy.*

Baby-splitting

Feeding fledged chicks is a demanding task for most birds. In the case of a lesser spotted woodpecker, a good breeding season may result in six healthy (and hungry) young leaving the nest at the same time and, even having acquired feathered independence, these tyros still need provisioning for at least another eight days.

Woodpeckers, along with quite a number of other garden birds, have a pragmatic answer to this last, exhausting parental stage. Known as brood-splitting or brood-division, it consists simply of dividing the brood into two, so that the male has full responsibility for half the brood and the female for the rest. Once it takes place, it is quite a stern discipline. Any chick attempting to get food from its non-assigned parent is vigorously repelled.

Take your partners

A detailed study of this small species in Sweden unveiled a surprising trait, unusual in the family: female lesser spotted woodpeckers quite often have more than one mate. About 10 per cent of all individuals hold two or more male partners simultaneously, both of whom help to feed the nestlings.

The advantage of such an arrangement was starkly clear. Overall, 34 per cent of breeding attempts by lesser spotted woodpeckers fail to produce any fledglings at all. However, the bigamous (or polyandrous) females are more successful than their monogamous counterparts: the number of young they raise is no less than 40 per cent higher.

Barn swallow

Species: ***Hirundo rustica***
Family: Hirundinidae

△ *Swallows are easily distinguished from house martins by their long tail-streamers. These birds often gather on wires prior to their migration.*

▷ *The large, very sensitive mouth enables the swallow to catch insects in flight.*

IDENTIFICATION The barn swallow (17–19 cm/ 6¾–7½ in, sparrow-sized but longer-bodied) is a common aerial bird of the countryside. It is usually seen flying low over fields and around farm buildings, but also perches on wires. It is identified by its long tail-streamers that trail behind it, by its deep burgundy throat and forehead, and by its iridescent blue upper parts and chest-band. The underparts are off-white. From below there is a white bar on the underside of the tail, which shows as a series of dots from above.

The *male* on average has longer and narrower tail-streamers than the *female*.

The *juvenile* (early summer to mid-autumn) is similar, but has much shorter tail-streamers. Its throat and forehead are brownish-pink, not red.

SHAPE AND CHARACTER The barn swallow has a long, streamlined body, swept-back wings and, in particular, its long, forked tail is highly distinctive, longer than in similar species. This effervescent bird flies purposefully, usually low over fields at some pace. It slips to one side or another with grace, and also glides higher up. It has a distinctive, strong 'rowing' action with its wings.

VOICE Its main calls have a twittering tone, but its alarm call is a distinctive 'witt-witt!' with an exclamatory tone. The males have a pleasing, rambling, twittering song, often interspersed with a dry, flatulent rattle (the more rattles that are included, the more dominant the bird). It sings in flight or from a perch, such as a wire.

HABITAT This is mainly a bird of rural areas, such as farms and fields, and also villages and smaller towns. It requires outbuildings or other structures for nesting, and also usually lives near water.

FOOD It eats flying insects, especially larger ones such as blowflies and hoverflies, taken in aerial pursuit. It takes larger items than the house martin or swift.

IN THE GARDEN The barn swallow often breeds in outbuildings with little encouragement, but you can also improvise a saucer-shaped base – even a half-coconut may work. Remember to leave a gap to allow the birds to enter.

BREEDING Both sexes build a nest inside an open building, or beneath a bridge or other construction. The nest is an open, shallow cup of mud mixed with vegetable fibres, placed on a support such as a rafter, and against a vertical surface. The female incubates four to five eggs for 17–19 days and both adults feed the chicks in the nest for 20–22 days, with parental care for some time afterwards. There are often two broods.

MIGRATION This bird is a migrant, arriving in early to mid-spring and leaving in early to mid-autumn (occasionally later). It winters in southern Africa, right down to the Cape of Good Hope.

ABUNDANCE It is a very common bird. In non-rural gardens it often flies overhead on migration.

BARN SWALLOW

CONFIDENTIAL

It's the tail

Experiments have shown that when it comes to pairing up, the most desirable male swallows are those with the longest and most symmetrical tail-streamers. Such birds are quickly selected by breeding females from the available pool, not just for the pair-bond, but for extra-pair liaisons as well. Scientists proved this by correlating tail length with breeding success, and by some rather unkind manipulation: when birds with good previous breeding success had their tails trimmed, their fortunes nosedived; meanwhile, males with short or asymmetrical tails that were given a little extra feather length, by means of glued-on tail-tips, soon reaped the benefits.

A long tail is a slight impediment to flying, so any male with extra tail length is showing that it can withstand these odds and still prosper. Furthermore, longer-tailed males have higher testosterone levels, suggesting that their immune system is in good working order. Individuals with asymmetrical tails have been shown to carry more parasites than their fit and equal-tailed rivals.

Tell-tale

Colonial swallows exist in a climate where, despite birds forming conventional monogamous pair-bonds, birds of both sexes routinely copulate with other individuals outside this arrangement. Such activity is rather helpful to a female, which can draw from a wide pool of quality to ensure that at least some of its offspring benefit from healthy genes.

To a male, however, extra-pair copulation by its mate threatens its paternity, and for several weeks each summer males work furiously to keep watch on their mates to ensure that, wherever possible, no such opportunities exist. The crucial period is during the week or so when the female is laying each clutch of eggs; if a male can keep its female honest during this time, it will probably ensure that the resulting offspring are its own. That isn't easy – swallows are rapid, manoeuvrable fliers and during the long spring days the female can often find a time when the male's concentration wavers; when this happens, the deed is quickly done.

However, if a male swallow does lose its mate for a moment, it has a trick up its sleeve. It cries wolf. By uttering the alarm call given when a dangerous predator is spotted, the male can ensure that everyone in the colony coalesces into a flock that can collectively watch or mob the predator. Admittedly no predator has actually been spotted, but the alarm call brings the flighty female back into view, where it can be safely monitored.

◁ Adult swallows are easily recognized by their long tail-streamers. The length and symmetry of the tail varies between individuals.

▽ In contrast to house martins, swallows usually build their nests inside buildings and on a horizontal platform.

Killing nestlings

Recent research has cast a shadow over the swallow's reputation as a sunny species for sunny days. Occasionally, it seems, swallows commit infanticide.

The circumstances are typical of those for infanticide in other animals: a male with breeding pretensions destroys the existing offspring of another. In this case, a male swallow might first lose a mate during the breeding season. It then selects a desirable (or neighbouring) female and, while both pair members are away from the nest, steals up to the nest, picks up a nestling in its bill and drops it some distance away – to certain death. It then removes the rest of the brood in the same manner. The loss of nestlings seems to result in splitting up the existing pair. The murderous male then frequently pairs up with the recently bereaved female.

Flight times

Swallows spend a large part of the year migrating. Leaving Europe in September, they arrive on the wintering grounds in South Africa in December. After a quick stay, they are off north by late February, and many don't arrive back in Europe until April.

House martin

Species: *Delichon urbica*
Family: Hirundinidae

IDENTIFICATION This small, highly aerial bird (12–13 cm/4¾–5 in) is well known for nesting in small colonies under the eaves of houses. It is similar to the swallow (see pages 40–43), but has a shorter, notched tail and a distinctive white rump. It has clean, snowy-white underparts, except for the dusky underwings. Above it is a dark, metallic blue-black, with dark-brown wings. The legs are feathered and white – a bird wearing 'long johns'.

The *male* and *female* look alike.

The *juvenile* (early summer to mid-autumn) is similar, but browner on the crown and with a dusky throat and flanks.

SHAPE AND CHARACTER The house martin is a highly aerial species, with a streamlined body and long wings, but nevertheless it is not as streamlined as the swift or swallow. It has a fairly short, notched tail. The wings are shorter and stubbier than the swallow's. It does not swoop as much as the swallow, but flies high in arcs, with much twisting and turning and semicircular glides. The wing-beats have a fluttering style, without any impression of power. It tends to feed higher up than the swallow.

VOICE Its main call is a hard, rolling 'prrit'. At the nest it makes a pleasing twitter based on the call.

HABITAT Ancestral house martins nested on cliffs, but this bird now relies on buildings for its nest sites. A town and village bird, it feeds aerially, often quite high. It needs mud and water nearby.

FOOD It takes small flying insects such as aphids and flies, caught in pursuit flight. It usually strikes from below, using the sky as a background.

IN THE GARDEN It is possible to buy artificial house martin nests to encourage them to nest in new areas. Provide several, close to the eaves. In hot weather, try to provide a mud puddle.

BREEDING Both sexes construct a half-cup of mud pellets, fixed to a vertical surface and flush to an overhang. They usually breed in small colonies and the nests may be right next to each other. Both sexes incubate four to five eggs for 13–19 days and the young remain in the nest until 19–25 days after hatching. This species has two to three broods, and the young can still be seen in the nest in early autumn.

MIGRATION A summer visitor, the house martin arrives in mid-spring and leaves in early to mid-autumn. It winters in sub-Saharan Africa.

ABUNDANCE The house martin is a common bird, but widely declining.

◁ *A group of house martins on a wire. Note how much shorter their tails are compared to swallows.*

▷ *A house martin collects mud for its nest. It may use up to 1,500 mud pellets to make a single structure.*

HOUSE MARTIN CONFIDENTIAL

Winter disappearance

The house martin is the eighth most numerous of Europe's long-distance bird migrants to pass over the Mediterranean and the Sahara each year on its way to tropical Africa, with more than 70 million individuals estimated to make the trip in any given autumn. It is therefore extraordinary that the house martin's main wintering grounds are still not known. It seems that these familiar European birds vanish into thin air.

Of the 300,000 house martins ringed in Britain to date, only a hundred have been recovered outside Britain; of these, only a single record comes from the wintering area in Nigeria. More than a million have been ringed in Europe and North Africa, but only 20 have given away their whereabouts south of the Equator.

There are a few sight records from the wintering areas and, interestingly, some are associated with flocks being driven down by bad weather. Furthermore, in Zambia most records are of birds flying very high up, in small flocks almost out of sight. This suggests that, after leaving their breeding areas, these birds might all fly up high in the air, never having to touch down. Somewhere in the vastness of Africa, therefore, the skies are full of house martins.

Night-time disappearance

Mystery also surrounds the night-time habits of house martins. It is known that they roost in the nest at night if they are breeding, but what do they do if they are not breeding? Swallows roost in reedbeds at night, but a search for house martins here returns a blank. So where do they go?

Some undoubtedly roost in the treetops, often in groups. House martins go to bed very late, well after dark, so they are not often seen going up to the trees, if they do this

▽ Most house martin courtship takes place at the nest site.

△ *These days it can be difficult to find house martin nests that aren't in a building, but in prehistoric times most were probably on cliffs.*

widely. There is also circumstantial evidence that they sometimes sleep on the wing. They can be grounded by bad weather at night, but no definitive picture of where they roost has yet emerged.

One thing proven is that they frequently use 'hotels'. House martins on migration routinely visit the nests of more southerly colonies as nightly stopovers, even when some of these nests are already occupied. However, there simply aren't enough nests to go round for every travelling house martin.

Nest pellets

The house martin's nest is a remarkable structure, made up entirely of mud pellets and stuck to a vertical wall. Many house martins, especially experienced adults, try to reuse old nests in any given year, because the consequences of not doing so mean a lot of hard labour. In many cases a nest takes two weeks to build – house martins usually have a session in the morning, then in the afternoon the mud can dry.

Once the foundation – a few pellets attached to a wall to make a shallow ledge – is in place, the birds build up and out. Every time they add a pellet they have to make sure no air pockets are caught between pellets, undermining the structure. The sheer effort of going to and from a mud puddle to a nest under construction is illustrated by the numbers: every nest incorporates 700–1,500 pellets, so even if mud is only 100 m (330 ft) away, that's still a good 300 km (185 miles) of commuting in the latter case. The construction can also be affected by the weather. Dry conditions can mean that suitable mud becomes temporarily unavailable.

Pied/white wagtail

Species: *Motacilla alba*
Family: Motacillidae

IDENTIFICATION This distinctive long-tailed bird (18 cm/7 in, slightly larger than a sparrow) walks on the ground rather than hopping. It is tame and often seen on lawns and roofs, wagging its tail. Its plumage is a mixture of black, white and grey. It has black legs and its long tail has white outer feathers. The wing feathers are black with broad white fringes. The head is mainly white, so that the black eye is obvious. In Britain and Ireland it is known as the 'pied wagtail' and in continental Europe a very similar bird is the 'white wagtail': their similarities and differences are described below.

The *adult male breeding* (early spring to late summer) pied wagtail lacks grey, and has black on the throat, chest and all the upper parts, including the back. The white wagtail has more white on the face and a neat pale-grey back, sharply demarcated from the black nape.

The *adult female breeding* pied wagtail is like the male, but the nape, scapulars (shoulders) and back are dark grey, not black. The white wagtail's back colour is like the male, but grades into a black nape.

The *adult winter* (early autumn to late winter) pied wagtail has black on the front, reducing to a band across the chest, with a white throat and belly. The white wagtail is paler, with clean pale underparts and a paler face.

The *juvenile* (late spring to late summer) pied wagtail's shape gives it away; its black is reduced to a chest band, brown-grey above, and it often shows a yellow wash to the face; it moults to first winter plumage in late summer, darker on the back, similar to the female. The white wagtail is similar; the first winter it resembles the female white wagtail.

SHAPE AND CHARACTER The long tail, together with the long legs that are used to walk and run along the ground, make this bird distinctive. When walking, it nods its head like a chicken and frequently wags its tail. It moves forward in fits and starts – walking alternating with darting runs.

VOICE Its main call is a loud, slightly spat-out 'twissi-vit', modified to a hard 'chissick' in flight. The song, not often heard, is a rambling medley interspersed with call-notes.

HABITAT The wagtail is found in a variety of places, often near water. It benefits from short turf and will even feed on roads and pavements, where it is easy to catch stray insects. It is often seen on roofs, but rarely settles on tree branches except to roost.

FOOD It feeds mainly on invertebrates, basically walking along and picking them up or chasing them. It also makes sallies into the air to snap at flying insects.

IN THE GARDEN This bird is not averse to visiting ground feeding stations, for crumbs left by other birds. It may take to an open-fronted nest-box, with 10 x 10 cm (4 x 4 in) inside dimensions, placed on a stone wall.

BREEDING It uses all kinds of nest sites, mainly crevices in walls, bridges, pipes, banks or the sides of buildings. Both sexes incubate five to six eggs for 12–14 days and, once the young hatch, they are fed in the nest for 13–16 days. It might take another two weeks for the young to become independent.

MIGRATION Some birds remain in the same area all year, whereas others move long distances south in the autumn and return in the spring.

ABUNDANCE It is a common bird.

△ *Wagtails will use a range of different sites for their nests. Cavities in walls and bridges are ideal.*

▷ *The white wagtail is a bold, common and approachable bird.*

PIED / WHITE WAGTAIL

Wagging doubts

Nobody seems to understand why wagtails wag their tails. Various theories have been put forward: that the birds are signalling to each other, reminding neighbours of their presence, or that the wagging may be a ruse to draw predators to the wagging bird while it is alert enough to avoid them.

A further theory suggests that the wagging helps to camouflage birds on the water's edge, enabling them to blend in with the flowing water and riverside vegetation. Perhaps the most persuasive theory, though, is that wagging flushes out insects, which the bird can then run and catch.

Roosts

Pied wagtails usually roost together at night, sometimes in their hundreds, often on roofs, but also in reedbeds, at sewage farms and in glasshouses. It is quite a regular sight to see them gathering on the roofs of superstores or garages. Occasionally they also gather in Christmas trees, evidently basking in the paltry heat of the Christmas lights.Nobody is sure why wagtails are drawn to each other's company at night, but it is an ingrained habit and they have a specific call that summons other birds to join them. It might be a defence mechanism against predation, or possibly the birds can exchange information.

Satellites

In the winter, pied wagtail society is divided. Some adult males hold feeding territories and defend them from other birds, while subordinates (including many females and juveniles) must make do with foraging in flocks – loose and often temporary associations. Individual territory-holding birds benefit from being able to run after prey with more abandon than flock members, since they don't risk stepping on another bird's personal space.

On occasions when there is plenty of food around, territorial males will tolerate a helper, known as a satellite. This subordinate bird helps the male with its boundary defence, at the same time sharing the plentiful resources within the territory. But although the arrangement may endure for a week or more, it is not lasting. If it gets colder, the satellite may be banished back to the flock while the territory-holder forages alone.

◁ *Modern science still hasn't worked out the full reason why wagtails wag their tails.*

▷ *Pied wagtails are in the habit of roosting communally at night, especially in the winter.*

Wren

Species: *Troglodytes troglodytes*
Family: Troglodytidae

IDENTIFICATION The wren (10 cm/4 in, much smaller than a blue tit) is one of Europe's smallest birds (the goldcrest is even tinier, see pages 88–89). It is a skulking but very noisy small brown bird, easily recognized by its short tail. It is mainly brown all over with fine, slightly darker barring, especially on the wings, tail, back and flanks. It has a long, slightly down-curved bill and a conspicuous pale eyebrow.

The *male* and *female* look alike.

The *juvenile* (late spring to mid-summer) is very similar, but lacks spots under the tail and is slightly mottled on the throat.

SHAPE AND CHARACTER The wren is a minute ball of feathers with rounded wings and a short tail, usually held cocked. The rather long, slightly curved bill is also distinctive. A very noisy and lively bird, it makes continual scolding movements and both flicks its wings constantly and curtseys on its perch; also curious, it will investigate unusual sounds (making a squeaking by sucking on your hand will entice it into view). Otherwise it is very hard to see, staying close to the ground in thick cover.

VOICE The wren's most common call is alarm, a loud, almost spitting, swearing 'teck!', often repeated endlessly. It also makes a rattle and a burst of calls like a machine gun. The song, remarkably vehement for the size of bird, is a long, loud, over-fast series of trills, usually including a loud buzzing in the middle of the phrase (imagine a sports commentator describing a sprint and you get the idea).

HABITAT It is not fussy, using almost any kind of dense, low vegetation – just the bottom 2 m (6½ ft), close to the ground, including hedges, herbs and thickets; it almost never strays higher. It often ventures into holes and other crevices, and also hunts on bark.

FOOD The wren explores within the tangle of stems and leaves above ground in low cover for a variety of small invertebrates, including many beetles and spiders, its main foods.

IN THE GARDEN To help wrens, ensure that the garden has a few untidy corners – wood-piles are excellent. They may come to the bird table for a few crumbs or grated cheese, and occasionally roost or nest in tit-boxes.

△ *Despite its size, the wren is a noisy and confident bird. Note how the tail is often cocked up.*

▷ *Wrens often feed close to the ground, amidst the tangle of vegetation.*

BREEDING The male begins nest-building (see pages 54–55) by making a domed structure in a hollow or cavity in the side of a tree, wall, creeper or bank. The female completes the nest and then lays five to eight eggs, which it incubates for 13–18 days. Once the young have hatched, they are fed by the female, with varying help from the male, until they leave the nest at 15–20 days of age.

MIGRATION The wren is mainly sedentary.

ABUNDANCE This is one of northern Europe's most abundant birds, a fixture in most gardens.

WREN CONFIDENTIAL

Cock-nests

The nest construction habits of the wren are far from conventional. Although both sexes build the nest structure, they don't do it at the same time; in fact, nest-building is central to the formation of pair-bonds.

Early in the breeding season a territory-holding male begins nest-building of its own accord, long before any female is on the scene. But it doesn't build just one nest – four and eight is normal, and up to ten exceptional. Furthermore, the nests are not complete; they are just the external shell and are not suitable for raising a brood of chicks.

This is where the female comes in. Initially attracted to the territory by the male's song, it proceeds to carry out an inspection of the male's nest-building efforts. The male is party to the checks and sings loudly as the female approaches the first structure, often while on top of the nest or even inside it. If the inspecting female enters any of the nests, the pair-bond is sealed and it will use the dwelling place for bringing up the young.

First, however, the structure must be completed. Using feathers and other light materials, the female lines the nest, 'furnishing' it with feathers. It can now be used for the rest of the breeding cycle. Meanwhile, with several other nests available, it is common for the male to attract a second female to its territory and therefore become polygamous.

Niche diets

In common with most garden birds, wrens sometimes attract attention with bizarre feeding techniques or unexpected food items. One often-quoted report concerns a wren that spent some minutes following a badger, picking up insects disturbed by the foraging and digging of the larger animal.

As for unusual foodstuffs, pride of place must go to an adult that nested beside a trout hatchery in England and fed its young almost entirely on fish fry. There are several other records of these tiny birds taking vertebrate food, including fish, young frogs and tadpoles.

◁ *A big mouthful for a small bird. Wrens feed mainly on insects, but they can turn their hand to unusual foodstuffs, too.*

◁ *Wrens are almost always very low down in the vegetation. They rarely if ever venture to the tops of even medium-sized trees.*

▽ *Pairing in wrens revolves around the male displaying its nest-building efforts to the female.*

Dormitories

On the whole, the highly strung wren is not a sociable bird. Males are exceedingly territorial and usually react aggressively in the close company of other birds. However, on occasion low temperatures force birds to put aside their differences and rely on each other to survive. On the coldest winter nights, they bed down together in dormitories.

A wren dormitory is often very small. A previously used wren nest, house martin cup or tit nest-box may suffice and, astonishingly, 60 individual wrens have been observed entering and leaving one tit-box at dusk and dawn. Typically, a single wren seems to be the 'ringmaster' of the roost, singing loudly to summon clients, but also acting as doorman and keeping visitors out, or even evicting them.

Conditions inside the nest are often uncomfortable. At a big roost the birds may actually squat on top of each other, in two or three layers, bills facing inwards. However, without sharing the body heat of other birds, the alternative may be much worse. In bad conditions some wrens, which are hardly great travellers, may commute 2 km (1¼ miles) to find a traditional site.

Waxwing

Species: *Bombycilla garrulous*
Family: Bombycillidae

△ *The exotic-looking waxwing is a rare winter visitor to Europe south of Scandinavia. This is a young bird.*

▷ *Waxwings are regular visitors to gardens, often taking advantage of thrown-out apples.*

IDENTIFICATION An exotic-looking bird (18 cm/7 in, smaller than a starling), the waxwing is unmistakable when seen clearly. It is a larger-than-average songbird with an obvious crest, pink-hued plumage, black face mask and chin and boldly patterned wings (black, with white and yellow stripes). The tail-tip is brilliant yellow, the vent is chestnut and the rump is grey.

The *male* has a well-demarcated black throat, broad yellow tail-band and the most patterned wing, with white curved edges to the tips of its flight feathers and waxy red tips to its secondaries (flight feathers attached to the 'elbow').

The *female* differs by having a diffuse greyish base to its black throat, less pronounced (or absent) white crescents on the wing and a thinner tail-band.

The *first winter* (early autumn to mid-spring) – of either sex – lacks crescents on the wing, so it just has a straight yellow line going down the wing; the red tips are reduced and the yellow tail-band is narrow.

SHAPE AND CHARACTER The crest is usually enough to identify the species. The waxwing is very sociable and is usually seen in flocks ranging from a few birds to hundreds. Quite mobile, it flies strongly with rapid wing-beats and a straight course; flocks keep very close together. The waxwing is agile in trees and can be nervous; but it is often extraordinarily tame, feeding within a few metres of crowds or traffic.

VOICE Its most obvious call is a ringing, silvery 'sreee', usually heard in flight.

HABITAT In the breeding season waxwings are found in remote, far northern coniferous forests, but then disperse anywhere that provides an ample supply of berries, including parks, roadsides, gardens and hedgerows.

FOOD It is unusual in taking large quantities of berries year-round, although it also takes a few insects at any time of year (especially midges in summer) by snapping them up in flycatcher-like aerial sallies. Sometimes it will 'catch' snowflakes in this way to drink. Its most important fruit is the rowan, with many others in the wild, including haws and whitebeam. It feeds in trees.

IN THE GARDEN Berry-bearing bushes, including cotoneaster, pyracantha and rowan, will attract and provide for visiting waxwings.

BREEDING This bird is very unlikely to be found breeding in any garden. It selects a mature conifer and builds its nest on a branch as much as 15 m (49 ft) above ground, usually close to the trunk. The waxwing lays five to six eggs, incubated for 14–15 days. The young leave at 14–15 days, having been fed on regurgitated berries and some insects.

MIGRATION The waxwing is a far northerly breeder that comes as far south as Britain, the Low Countries and the Balkans to winter, but its regularity decreases the further south you go. In some years of poor berry crops it is found more widely, and in much greater numbers than usual.

ABUNDANCE This bird is usually uncommon.

A berry diet

Waxwings are the only birds in Europe that are specialist frugivores (fruit-eaters). That means that they can eat nothing but fruit and still survive, whereas other birds that eat fruit, such as thrushes, must also eat insects to obtain the nutrients they need. At many times of the year the waxwing's diet consists of 100 per cent fruit, mainly in the form of berries. It even feeds fruit to its young in the nest, although for the first two days they are given insects.

A waxwing's consumption of berries is pretty impressive. On a normal day it eats 600–1,000 of them, equivalent to more than twice its body weight. In order to harvest them effectively, the waxwing has a broad gape (mouth) 1.1 cm ($\frac{4}{10}$ in) wide, as big as that of a fieldfare.

Irruptions

Waxwings are famous for their periodic 'invasions' into areas where they are not normally found, including Britain. Without warning, the birds pour south in early winter and suddenly, much to the delight of birdwatchers, can be found everywhere, often – bizarrely – in superstore car parks where berry bushes are planted. These 'irruptions', as they are properly called, last the whole winter, with the birds returning north in early spring.

Nobody has definitely explained why these occasional events (once a decade or so) occur. However, it is probable that they are a result of high population caused by a good breeding season, together with a major failure in the rowan-berry crop on which they usually depend.

Dunnock

Species: ***Prunella modularis***
Family: Prunellidae

◁ *Often confused with a house sparrow, the dunnock is similarly brown-streaked but has a greyish head, reddish eye, pink legs and a much thinner bill.*

▷ *The dunnock is primarily a ground feeder and will often creep about on the grass below bird tables. Look out for its wing- and tail-flicking.*

IDENTIFICATION The dunnock (14.5 cm/5¾ in, the size of a robin) is a retiring, easily overlooked small brown bird, with a thin insect-eater's bill. It is usually seen on the ground. Rather sparrow-like, with brown, streaky plumage, its head and neck are suffused with grey. The brown plumage is strongly streaked with darker brown on the upper and underparts. The wing has very narrow whitish wing-bars. The eye is red.

The *male* and *female* look alike.

The *juvenile* (late spring to late summer) is much more strongly streaked than the adult, especially below, and has only a little grey on the head and neck.

SHAPE AND CHARACTER The thin bill distinguishes it from similar seed-eating birds. It may be small and brown, but the dunnock has a manner all its own. Often seen on the ground, it seems to have a 'shuffling' or 'creeping' gait, moving forward with its legs flexed in a crouching style. It constantly flicks its wings and flirts its tail nervously. It keeps to cover and is unobtrusive. It is not sociable.

VOICE The dunnock's most commonly heard call is a piping 'seep', often quite loud. Its song (mid-winter to mid-summer) is a fast, high-pitched, cyclical warble, which has been compared to the sound made by the unoiled wheels of a trolley.

HABITAT This is a bird of low scrub. In gardens it feeds among leaf-litter, on the edges of lawns and on pathways. It uses higher perches, such as fences and trees, to sing or call.

FOOD It chiefly eats very small invertebrates, including spiders, beetles, springtails and tiny snails. It also takes small seeds, especially in winter. Its feeding method is simply to pick stuff up from the ground.

IN THE GARDEN It will come readily to ground stations for various crumbs, small seeds (including lentils), grated cheese and soft-bill mixes.

BREEDING The female builds a fairly substantial cup-nest, placed in a bush, low hedge or tree, well concealed. It lays four to five unusual bright-blue eggs and incubates them for 12–13 days. The young are fed by the female, with additional help from one or more males (see pages 60–61), and leave the nest after 11–12 days; they may be tended for up to 17 further days.

MIGRATION The dunnock is resident in Britain and nearby Europe, but is a summer visitor to Scandinavia.

ABUNDANCE It is common in Britain, local in continental Europe.

△ *Dunnocks often have the habit of singing about half a dozen times from a song post and then moving on to another one.*

Fancy a quickie

Despite the tangled web it weaves, the actual act of copulation in the dunnock is remarkably brief – the cloacal contact lasts for just a fraction of a second.

Nevertheless, the dunnock exhibits a unique pre-copulation display. The female droops its wings and quivers its tail, while the male hops behind it. At this, the male pecks at the female's cloaca, which is distended, about 30 times. After a minute or so the cloaca ejects sperm from the previous copulation. This is evidently to assure a male that its own sperm is now favourite to fertilize the next egg-to-be.

'Alpha' males and 'beta' males

Anyone researching dunnocks soon finds that little in their social lives is simple. For example, in contrast to most other species of small birds – in which a male sets up a territory and sings to defend it – it is female dunnocks that are territorial and sing early in the breeding season. Mating is initially effected by a male 'taking over' the defence of a female's territory. If the male is capable of defending it, the pair may live as a monogamous couple.

However, in many cases a female territory is too big for a male to defend on its own (females have a lower population level, and therefore more 'room', than males). In this case the original male may be forced to let in another male to help it keep the borders intact. When this happens the territory may be shared between two males and a female, but the tension is palpable. The first (or 'alpha') male is dominant, and tries to keep access to the female exclusive to itself; if it sees the subordinate ('beta') male approaching the female, the alpha male will drive it off.

However, while the alpha male drives the beta male away, the female is complicit in exactly the opposite happening: it and the beta male can often be seen copulating literally behind the alpha male's back. The alpha male's paternity is compromised, but the beta male is satisfied and the female has an extra helper on its list.

▷ *The remarkable display of dunnocks about to copulate. The male repeatedly pecks the female's cloaca to cause the ejection of sperm that might have come from another male.*

Mate-sharing

In recent years the dunnock has acquired cult status among biologists for its extraordinary breeding activities. Somehow, a tendency towards multiple mates seems at odds with its image as an inoffensive, retiring, small brown bird.

What makes dunnocks unusual is that they will form pair-bonds with several members of the opposite sex. There are usually more males than females in a population, so it is routine for a female to have a relationship with two males, and sometimes three or even four. Males often have just a single female mate; sometimes two, rarely three. However, these relationships may overlap, so that a male paired with two females might find itself sharing both of them with another bird, and sometimes with two. And so on.

At first this sounds like simple promiscuity, but it isn't. The central tenet is that any male that mates with a certain female is effectively promising to help it feed the subsequent brood. Dunnocks feed minute insects to their young and need all the help they can get. Hence it is the female's desire to enlist helpers, by allowing them to copulate with it, that drives the mate-sharing system.

▷ *The thin bill of the dunnock only enables it to gather very small food items.*

Robin

Species: *Erithacus rubecula*
Family: Muscicapidae

IDENTIFICATION This familiar garden bird (14 cm/ 5½ in) has an orange breast; the orange extends onto the forehead, making a 'double-breast' of two lobes of colour with white indentation in between. It is white on the belly. The upper parts are warm brown except for a narrow line of grey bordering orange on the forehead and cheek.

The *male* and *female* look alike.

The *juvenile* (late spring to early autumn) breast may show a warm, red-brown wash; otherwise it is just brown with pale spots on the upper parts, a more complicated pattern on the underparts, with dark chevrons surrounding the spots.

SHAPE AND CHARACTER This small, brown bird is very common and easy to see. It feeds on the ground, moving by hops, and doesn't run like a thrush. It often perches low down, for example on a spade. As a wild, woodland bird, it lives in the shade and can be very retiring.

VOICE Its call is a distinctive, metallic 'tick', similar to the sound of a car engine cooling down. Its song is slow and melodic, but somewhat wistful and shrill. The pattern is for the bird to sing a phrase, leave a gap for a response, then sing a completely different phrase. Uniquely, the song is uttered all year round except for a break in mid- to late summer, so it is the only garden bird likely to be singing much in early autumn.

HABITAT In the garden, all a robin needs is low shrubbery where it can roost and nest, and some grass/lawn, litter or uncluttered ground over which to feed. Otherwise it is not fussy.

FOOD In the wild it is mainly insectivorous, feeding on beetles and spiders and other fast-moving creatures. It also takes small fruits and berries in autumn and winter, and occasional seeds.

IN THE GARDEN Robins come readily to bird tables, even to hanging bird feeders. They are specially fond of mealworms, but also take cheese and soft-bill mixtures. They will use a standard open-fronted nest-box. Place it low down, up to about 1 m (3⅓ ft) above ground, among vegetation such as a creeper.

BREEDING The robin breeds at any time between early spring and mid-summer. The female builds a cup-shaped nest, often in a hollow or very low on a tree stump. It lays four to five eggs, which are incubated by the female, fed regularly by the male, for 15 days. The young are fed by both parents and leave the nest after 13 days. There are two or three broods in a season.

MIGRATION Most robins are resident, but some (especially females or young) migrate towards Spain or Portugal from early autumn, arriving back in early spring. A few come from north-eastern Europe for the winter. When they do migrate, robins move at night and are known to use the stars for orientation.

ABUNDANCE This bird is very common.

△ *The juvenile robin, with its spots and ochre-coloured breast, bears little resemblance to its parents.*

▷ *The usual feeding method of robins is to find an elevated perch and watch for movement below.*

ROBIN
CONFIDENTIAL

The missing robin
Although a mainstay of the garden scene throughout the year, the robin is often difficult to see in mid- to late summer, and householders often panic that 'their' robin has come to grief. A more likely explanation is that the bird is keeping a low profile during the energy-sapping moult.

The night singer
Robins often sing at night and, in the past, this has led garden watchers to wonder if they are hosting nightingales on their patch. On many occasions, especially in early to mid-winter, it seems incongruous to hear birdsong in the darkness. Recent studies suggest that robins sing at night because it is a quiet time of day, when their message can be transmitted more efficiently than during the hours of light.

Fierce robins
Robins are famously aggressive towards their own kind, and it is quite regular for fights to lead to the death of a combatant if rivals are well matched. The fiercest battles are in the autumn and early spring, and are inevitably over territory. Quite simply, a place to live and feed is everything to a robin, so they are battling over high stakes.

When one robin challenges another, it simply trespasses into its territory and makes 'tick' calls or sings. This elicits a reply in kind from the territory-holder. For a while there will be a battle of words, with the song phrases becoming increasingly hurried and strangled. If nothing is resolved, the birds will face up to each other, showing the orange breast and manoeuvring the body so that this feature is obvious to its rival. Only after much display and singing will a fight break out.

Actual fights are shockingly fierce and often brief. The birds peck and claw at each other, aiming at the head of their rival. A good strike is often enough for a kill.

Feeding style
The robin's feeding style is very much 'watch and wait'. Either it stands in the grass or litter waiting to detect food at its feet, or it finds an elevated perch to watch the comings and goings of invertebrates below. If it spots an invertebrate out in the open, it will fly down to snap it up. It so happens that spades make ideal perches for this method of feeding. This may also explain why robins are so fiercely territorial. Watch-and-wait doesn't work if the viewed area is constantly being disturbed, so it pays to keep competing robins away. Other trespassers, particularly dunnocks (and even wood mice), may also be shooed away.

Unusual nest sites
Robins are famous for their frequent choice of unusual nest site. Strange places have included coat pockets, skeletons, drawers in filing cabinets, hats and flowerpots – even an unmade bed.

◁ *A stand-off in your garden. Robins will ward off other birds from their watching perches to ensure a regular feed.*

Tame robins

△ *The gardener's 'friend'. Robins will often wait for people to turn over the soil in expectation of an easy meal.*

Robins are famous for their apparent ease with humans, coming up to us in an enquiring way and demanding food, often singing as they do so. Very few birds are so tame.

The habit may arise from an ancient method of finding food in the robin's ancestral forests, where animals such as wild boar, moles or deer frequently disturbed the soil when feeding or wallowing, turning up previously hidden invertebrates that the robin could feed on if it followed the mammal's exploits. It was then only a short step to doing the same to gardeners.

▷ *Battles between robins are swift, but deadly. Each will strive for a position from which it can inflict a blow from its beak.*

Blackbird

Species: *Turdus merula*
Family: Turdidae

△ *Female blackbirds are brown. Take care not to confuse the speckled pattern on the breast with that of a song thrush.*

▷ *Blackbirds eat large numbers of berries, especially in the autumn. Note that this male is a youngster in its first winter, with a black bill.*

IDENTIFICATION This common, medium-sized songbird (24–25 cm/9½–10 in) can be seen feeding on lawns or perching in high places to sing in the spring. It has the thrush-like features of large eyes and a longish tail.

The *male* is unmistakable, sooty black all over except for a bright orange-yellow bill and eye-ring.

The *female* is also plain-coloured, but more a dull dark brown, slightly paler below, especially on the throat and upper breast, often with dark-brown streaks on the chest. The bill is dull yellow. Some females are lighter brown than others.

The *juvenile* (mid-spring to early autumn) is a warmer chestnut-brown than the female, and speckled with dark spots below and pale spots on the mantle. The bill is dark.

The *first winter male* (mid-autumn to early winter) is as the adult male, but the bill is black until the end of the year.

SHAPE AND CHARACTER The blackbird is a robust songbird with strong legs, a longish tail and a quite long, strong bill. It is often seen feeding over lawns, running along and then pausing, motionless, looking and listening for invertebrates in the soil. It is not very sociable and quite aggressive. When landing, it often lifts its tail and then brings it slowly down again. It is prone to sing from high perches, such as rooftops and tall trees.

VOICE It is noisy, with several calls, including a panicky alarm-rattle when flushed, plus a soft 'chook-chook' and an annoyed, repetitive 'chink! chink!' at dusk or when angry. Its gorgeous melodic song (late winter to mid-summer) is rich and flutey. It is uttered in phrases, different each time, starting well, but descending into a chuckle, with pauses between each.

HABITAT A woodland bird at heart, the blackbird does well in the suburban mix of lawns, trees and shrubs. It forages over lawns and in flowerbeds, feeds on berries in trees and shrubs, and nests in hedges.

FOOD For much of the year it feeds on invertebrates in the soil, especially worms, obtained by watching for their signs on the ground. It also eats caterpillars, beetles, snails and even small fish and newts. In autumn and winter it eats berries in large quantities.

IN THE GARDEN This is a 'natural' garden bird that comes to flat feeding stations or the ground for scraps such as cheese, sultanas, apples and seeds. It will sometimes take to open-fronted nest-boxes.

BREEDING It may start breeding in early spring. The female builds a cup-nest and incubates three to five light-blue eggs alone, for 11–17 days. Both sexes feed the young on worms until they leave the nest at 12–19 days, and parental labour may continue for another three weeks. There may be three broods in a season, very occasionally more.

MIGRATION In much of Europe the blackbird is resident, but birds from Scandinavia and the east migrate south and west in mid- to late autumn to spend the winter in Britain southwards, returning in early spring.

ABUNDANCE It is extremely common.

BLACKBIRD CONFIDENTIAL

△ *The fabulous song of the blackbird is a seasonal treat: it is heard between February and July only.*

Growing tuneful with age

In blackbird society there is no doubt that a singer's ability improves with age. Mature singers are the best – no argument. This isn't just a reflection of the greater experience of older birds; there's an empirical difference, too. A singer's repertoire becomes wider and richer with age. In many bird species, all song learning takes place within the first year of an individual's life, but in blackbirds it seems to continue year-on-year. At any rate, the songs of older birds contain more elements, in more combinations, with less repetition, than those of birds at the outset of their singing career.

Conditioning the blackbird way

Some types of bird behaviour are easier to see in the garden than anywhere else, and an excellent example is plumage care. Blackbirds, in particular, show off interesting ways of looking after their feathers.

One such behaviour is sunning. The blackbird finds a sunny corner, spreads its wings and tail, ruffles its plumage and, often, opens the bill in what looks like an ecstatic pose. Exposure to strong sunlight probably helps to disturb parasites and enable the bird to pick them off; it is good for feather structure; and when preen-oil is exposed to the sun's rays, vitamin D is automatically produced.

A similar ground posture may reveal a more unusual behaviour, known as 'anting', in which the bird allows ants to crawl all over its body. Here, too, the insects may help to remove parasites from the bird's plumage.

Tension across the lawn

When blackbirds feed, it is a recipe for tension. Ground foraging is not an easy, sociable activity. Birds have to keep a personal space between them, because finding food requires careful searching – and listening – for small clues on the surface. Somewhere just under the grass is a big reward in the form of a worm, and adjacent birds are automatically in competition for it. That is why tension is inevitable when two birds share, for instance, a patch of lawn. It might not result in physical conflict; the birds may even seem to ignore each other. But the unspoken awkwardness tells in the feeding rate – when one blackbird is joined by another, its feeding rate drops by 60–80 per cent.

◁ *The colour of the bill of a male blackbird is a strong clue to its state of health: the more orange, the better.*

A bill is the price you pay

We can recognize an adult male blackbird by its orange-coloured bill. And it turns out that so do female blackbirds – and in a much more intimate way.

Bill colour varies between individuals. Some blackbirds have intensely orange bills, while others are a duller yellow. But this colour is not fixed for long periods of time. An orange hue can quickly change to yellow (over a matter of days) and vice versa. It's all to do with the health of the bird.

The reason lies in the pigments that colour the bill orange – the carotenoids – which are found in the birds' food and also play a part in its immune system. When a bird is fit and healthy, the carotenoids both help to fight disease and colour the bill yellow. A less healthy bird needs all its carotenoids for immune defence and cannot spare any for unnecessary bill coloration.

The bill is therefore an honest reflection of a bird's overall health. It isn't something that can be faked. Not surprisingly, very orange-billed birds enjoy better breeding success, mainly because they tend to pair up with healthier females. The females can see when a bird is 'off colour', in every sense.

◁ *A blackbird sunning. The bird ruffles its feathers in order to expose as much skin as possible to the rays of the sun.*

Fieldfare

Species: *Turdus pilaris*
Family: Turdidae

IDENTIFICATION The fieldfare (25–27 cm/10–10½ in) is about the size of a blackbird, but looks larger and stands more erect. It is a sociable thrush with smart, distinctive plumage. It has an ash-grey head, nape and lower back, strongly contrasting with the black tail and velvety red-brown band across the back. The underparts are whitish, decorated by streaks and chevrons, with a strong stain of ochre on the upper breast. A white patch at the shoulder is often obvious at a distance.

The *male* and *female* look alike.

The *juvenile* (mid- to late summer) is much duller and browner, with whitish streaks on the mantle and black spots on the underparts; it moults into first winter plumage by early autumn, when it is distinguished from the adult by pale tips to its greater wing-coverts.

SHAPE AND CHARACTER This is a large, upright-standing thrush with a relatively long tail. It often feeds far from cover, in the middle of a field, as well as closer to cover. It has a distinctive flight, with fast bursts of its wing-flaps followed by short glides, and often flies high. It is sociable, often seen flying in flocks, which are undisciplined and loose. On fields, the flocks move across from one side to the other, with each bird stop-starting; the whole manoeuvre can look like an army moving cautiously across a space.

VOICE Its main call is a noisy 'shack-shack', often given by flying birds. The song is a discordant mixture of chatters.

HABITAT The fieldfare breeds on continental Europe on the edges of woodlands, sometimes in parks and gardens. In winter (including in Britain) it occurs in similar places, often entering gardens to plunder berries from shrubs.

FOOD It eats invertebrates in the breeding season, including worms. In winter it starts off by feeding mainly on berries, but as the winter progresses turns back to worms and insects. It can dig through snow and into the soil to feed.

IN THE GARDEN It usually only visits gardens in hard weather when the ground is frozen, taking thrown-out apples, berries on bushes and even fruit laid out on bird tables.

BREEDING The fieldfare breeds in continental Europe, especially Scandinavia, but not in Britain. It builds a bulky cup-nest, placed in the fork of a tree. It is often loosely colonial. It lays five to six eggs, which the female incubates for 11–14 days; the young leave the nest 12–16 days after hatching and are fed by both parents.

MIGRATION Breeding birds leave in early to mid-autumn to winter at varying distances west and south. Scandinavian birds arrive in Britain in large numbers for the winter, with some returning as early as late winter.

ABUNDANCE This bird is fairly common.

◁ *The grey tones on the fieldfare's head and rump easily distinguish it from other thrushes. The narrow white bar across this bird's wing ages it as a first winter, a few months old.*

FIELDFARE CONFIDENTIAL

△ *A complete set of European thrushes at a single feeding station. Clockwise from top: fieldfare, song thrush, blackbird (male), redwing, mistle thrush.*

▷ *Predators such as this hooded crow risk bombardment with faeces if they come too close to a group of nesting fieldfares.*

Nest protection

The fieldfare is justifiably famous for its remarkable way of dealing with dangerous intruders close to the nest. The protection is so effective that a whole suite of other birds, including redwings and bramblings, regularly nest within the zone covered by a fieldfare 'colony'. Fieldfares tend to nest in reasonably close proximity to each other, the nests being about 10 m (33 ft) apart, with tens or hundreds within the same vicinity.

By having many pairs close at hand, fieldfare defence can be collective, with dozens of birds attacking a potential predator at once. However, these birds' actual technique in repelling threats is quite unique. Should a predator get too close, a variable number of birds will rise up from their nest sites, give a discordant call, dive-bomb the intruder and, remarkably, aim a dollop of excrement at it.

To a feathered predator, a matting of bird excreta can be dangerous. If pelted by too much, its plumage could be seriously damaged and its waterproofing possibly impaired, and this could be fatal. No wonder so many predatory birds give fieldfares a wide berth.

Benefits of mixed flocks

Fieldfares are more successful at feeding on old pastureland when redwings accompany them. They have a higher intake of energy in mixed flocks, as opposed to flocks of just their own species. It seems that they watch and learn about the whereabouts of prey from their smaller relatives.

However, fieldfares are better at searching newer pasture that is less than four years old. Redwings usually avoid this habitat, because there are fewer worms, but fieldfares apparently find what worms there are easier to detect in this habitat.

Mistle thrush

Species: *Turdus viscivorus*
Family: Turdidae

IDENTIFICATION This is a pale-brown thrush (27 cm/10½ in, larger than a blackbird) with a rather washed-out look. The breast is largely white, with big round spots, which sometimes seem to coalesce on the sides of the breast to make a dark patch. The white tips to the tail are distinctive, as is the pale panel on the wing, caused by pale edges to the flight feathers. There is a white patch on the cheek. The rump is paler than the back. The underwings are white.

The *male* and *female* look alike.

The *juvenile* (mid-spring to mid-summer) is similar, but has pale spots on the back.

SHAPE AND CHARACTER The mistle thrush has a distinctive shape, with a heavy, barrel-chested body, upright stance, long tail and oddly small head. It moves forward with heavy hops. It is not very sociable and can be highly aggressive. It is usually seen away from cover. It flies with a distinctive, heavy undulating flight, away to the tops of trees, not to cover, and sings from a high perch.

VOICE Its most common call is a dry rattle, which intensifies when the bird is alarmed. Its song is midway between a blackbird's and a song thrush's (but faster), with some repetition. It has a mournful, far-away ring. The phrases are short and, although the bird doesn't repeat them one after another, favourite phrases crop up.

HABITAT This is a bird of open woodland, requiring both tall trees for nesting and glades or pasture for feeding. It tends to feed far out in the field, a long way from cover. It really only occurs in large gardens, or those on the edges of fields.

FOOD As with other thrushes, it feeds on invertebrates (including worms) for much of the year, supplemented by berries in autumn and winter. It prefers certain berries, such as sloe, haw, holly and mistletoe. Animal food includes slugs, beetles and snails.

IN THE GARDEN It will come to feeding stations, especially on the ground, for fruit and sultanas. Berry-bearing bushes also attract it.

BREEDING The mistle thrush begins breeding early, often with eggs in early spring. It builds a large cup of grass, stems, roots, moss and dead leaves, often up to 9 m (30 ft) above ground in the fork of a tree. The clutch size is four to five eggs, incubated for 12–15 days by the female. Both sexes tend the chicks, which can fly when they are 20 days old.

MIGRATION It is resident except in Scandinavia, where it is a summer visitor. These northern birds evacuate their breeding grounds in mid-autumn. There may also be considerable local movements.

ABUNDANCE The mistle thrush is fairly common.

◁ *Not always easy to tell apart from the similar song thrush, the mistle thrush has a whiter background to its breast spots, a pale wing panel and a generally paler grey-brown colour to its plumage.*

▷ *Mistle thrushes rely on berries to supplement their diet in winter and will sometimes defend berry-bearing trees against other birds to protect their food supply.*

MISTLE THRUSH

◁ *In common with other thrushes, mistle thrushes eat a lot of berries. But unusually, they will often opt to defend a single crop on an individual tree or group of trees for a long period of time.*

◁ *A key identification feature of the mistle thrush is the white underwings. Note also its rather barrel-shaped body.*

Berry treasure

In the autumn and winter some mistle thrushes, usually acting as pairs, take over a berry tree or group of trees as their own, and attempt to defend it aggressively from all other berry-eating birds long term. The aim is to maintain the berry stocks – and they must be long-lasting berries such as hollies or ivy – as a personal winter food supply. The birds won't spend the whole winter eating them; when conditions are mild, they will switch to earthworms and other ground-living invertebrates. But even when feeding some distance away the owners will keep an eye on 'their' tree, making sure trespassers stay away.

When this resource protection works, the berries may last throughout the winter, even into spring and occasionally right into the breeding season (mistle thrushes have been seen feeding their winter berry stores to their young in the nest). However, disaster can strike if a large flock of hungry travelling birds, such as fieldfares or starlings, makes a stop at the allotted trees. On such occasions, with so many intruders present, not even the large and enraged mistle thrush can prevent the berry supply being stripped bare.

Singing in the rain

Mistle thrushes often sing at unusual times, when other birds are silent or their song is suppressed. For example, in the afternoons, when there is a widespread lull, mistle thrushes may be in full song. Similarly, these birds also sing in wet and windy weather, and at an approaching storm.

Raising the white feather

It so happens that the two most aggressive garden thrushes, the fieldfare and the mistle thrush, both have white undersides to the wing. Not surprisingly, there is a link. When either species is aggressive, it lifts its wing to expose the white in a flash of warning.

Song thrush

Species: *Turdus philomelos*
Family: Turdidae

IDENTIFICATION The song thrush (23 cm/9 in) is smaller and more furtive than the closely related blackbird. It is dark brown above, with pale underparts decorated by attractive dark spots. The spots are well spaced and often arrow-shaped. It is a pale sandy-brown under the wings.

The *male* and *female* are alike.

The *juvenile* (mid-spring to late summer) is similar to the adult, but has pale spots on the back.

SHAPE AND CHARACTER This is a compact, well-proportioned bird with a large head and medium-length tail. Like the blackbird, it typically runs along the ground, with regular stops to look and listen for food, often tilting its head slightly as if listening. It usually stays close to cover and is not seen far out in the middle of fields like the mistle thrush. When flushed, it quietly makes for cover, flying low.

VOICE When flushed, it gives a soft 'tick', as if 'tutting' in mild agitation. Its song (late autumn to late spring) is loud and dominant. It is very distinctive for its utterance of the same phrases (motifs) several times in succession before moving on to the next, and equally distinctive for its measured pace. It has better diction than the mistle thrush, but is not as tuneful as the blackbird's song. Blackbirds are often quiet during the song thrush's period of domination in early to mid-winter.

HABITAT This bird thrives on the tapestry of lawns and shrubbery in smaller gardens. It is easy to see when feeding on a lawn. On winter and spring mornings it sings loudly from high vantage points such as treetops and aerials.

FOOD Its main foods are invertebrates gathered from the litter or soil and, in season, fruits. It also likes snails.

IN THE GARDEN It will help song thrushes if you keep the garden free of pesticides and herbicides, to boost the snail population. This bird will visit ground stations sparingly for cheese, fat, sultanas and fruit.

BREEDING It may have eggs in the nest in early spring, and can bring up three broods in a season (and occasionally four). It builds a nest in a tree, shrub, creeper or hedge, usually close to the trunk. There are four to six sky-blue eggs, which are incubated by the female for 14–15 days. The young leave the nest 14–15 days after hatching.

MIGRATION Some individuals are sedentary in Britain and continental Europe, but birds also arrive in mid-autumn from Scandinavia to spend the winter. Others go as far south as North Africa.

ABUNDANCE The song thrush is a common bird.

◁ *The main food of the song thrush is invertebrates that live in the soil and leaf-litter, so it isn't difficult to tempt them to feeding stations with mealworms.*

▷ *A marvellous portrait of a song thrush. Note the well-proportioned shape, dark upper parts and dense, drop-shaped spots on the breast – all good distinctions from the mistle thrush.*

SONG THRUSH CONFIDENTIAL

Singing at dawn and dusk

Once a male song thrush has paired up and a breeding attempt is under way, it usually cuts down its singing radically, normally performing only at dawn and dusk. This means that, if you hear a song thrush during the day at this time, it probably is not paired.

If the cup fits

Studies of song thrushes have revealed that they undergo what can only be described as 'house-hunting'. Obviously it is important for any bird to have a suitable nest site, but what finally makes a bird settle upon one particular place is not readily understood. In the case of the song thrush, comfort would appear to be a factor. Females have been seen hopping one way and then another around a potential site, then squatting in the nest-to-be and even rotating their bodies within the 'cup' using their feet, keeping the tail up. Much the same takes place during the actual nest-building phase of many birds. It looks like the equivalent of going to a furniture shop and testing a range of sofas to check how comfortable they are by sitting on them.

Snails and anvils

Song thrushes are famous for their ability to smash snail shells to get the juicy molluscs inside. When a bird finds a snail, it carries it in the bill to a hard surface, such as a rock, tree root or wall, then pounds the shell on the surface until it breaks open. Surfaces that are profitable for this are used regularly and are known as anvils; they can be identified by their piles of shell fragments.

The technique of breaking the shell involves a sideways flick as well as a downward thrust. Surprisingly, the song thrush is alone in having mastered this, and blackbirds cannot seem to manage it. Even so, they are often attracted by the sound of the smashing and will sometimes steal the exposed snail at the last moment. The ability to eat snails is particularly useful when the ground is hard and other food isn't available.

◁ *The tapestry of short grass, hedges and trees that constitutes the suburban landscape is perfect for song thrushes, which commonly breed here.*

▷ *A song thrush smashes open a snail shell. No other thrush species are able to do this.*

Redwing

Species: *Turdus iliacus*
Family: Turdidae

IDENTIFICATION Smaller than other thrushes, the redwing (21 cm/8¼ in) is a distinctive thrush with stripes (not spots) on a white underside, red-brown flanks and a stripy face, with a pale-whitish supercilium (eyebrow) and a pale stripe from the bill down the side of the cheek.

The *male* and *female* are alike.

The *juvenile* (mid- to late summer) has pale spots on the back.

The *first winter* (early autumn to early spring) is similar to the adult, but often has face stripes stained buff and shows pale spots on the tertials (feathers on the inner wing next to the rump).

SHAPE AND CHARACTER This is the smallest of the typical thrushes, with the shortest tail. Otherwise it acts much like a song thrush (see pages 76–79), feeding over pasture and stop-starting as it hunts for soil invertebrates. However, it is much more sociable than the song thrush and often seen in large flocks. Its flight is only slightly undulating, and fast like a starling. Flocks keep together better than fieldfare flocks. It is often heard at night migrating or travelling low overhead (mid-autumn to early spring).

VOICE Its main call is a sharp 'seep', often heard as the bird flies overhead. In late winter/early spring flocks sit together and 'babble'. Its song (mid-spring to early summer) is a delightful short, descending phrase with a mournful feel, not usually heard in gardens.

HABITAT This is a bird of northern mixed woodland, also occurring in town parks and gardens. In winter it is found on woodland edges, and typically visits gardens for berries and fruit. It often forages in leaf-litter under trees.

FOOD Like other thrushes, it feeds on invertebrates at any time of year, with fruit predominating in autumn and winter, especially berries. Its foraging technique is a typical stop-start running: it stops to scan, moves forward, pecks at what is seen, and so on. It may also sweep its bill sideways in litter.

IN THE GARDEN The redwing is usually a hard-weather visitor to gardens, calling to take berries from cotoneasters, hollies and other trees and shrubs. In extreme weather it may visit feeding stations for fruit, including apples.

BREEDING It does not commonly breed in Britain or temperate continental Europe, but mainly in Scandinavia, Iceland and the Baltic eastwards. It nests in thick vegetation, often close to the trunk of a small tree, but sometimes on the ground. It usually brings up two clutches of four to six eggs, incubated for 12–13 days. The young fledge 10–15 days later, but are fed for at least two weeks afterwards.

MIGRATION The redwing is highly migratory. In most of Europe (including Britain) it is a winter visitor, arriving in mid-autumn and departing in mid-spring; it is often heard calling as it migrates at night. In Scandinavia it is chiefly a summer visitor, arriving in early and mid-spring.

ABUNDANCE This is a common bird.

◁ *It is the redwing's patterned face, with buff stripes over and beneath the eye, which most readily distinguishes it from other thrushes.*

▷ *With its brown back coloration and habit of freezing, the redwing excels at escaping the attentions of predators such as the sparrowhawk.*

REDWING CONFIDENTIAL

Winter visits

In many migratory bird species it is well known that individuals return to exactly the same breeding site each year after a winter away. In a good many species, individuals also show a similar attachment to the wintering site, coming back again and again to the same place from mid-autumn to early spring.

The redwing, however, is not one of these. Its attachment to a wintering site is zero – indeed, it is well known for spending successive winters in completely different places. For example, redwings ringed in Britain one winter have been recovered in Italy, Greece, Israel and even Russia the following winter – sometimes up to 5,000 km (3,100 miles) from the previous year's wintering site. Interestingly, even individuals raised in the same nest the previous summer may be spread far and wide in different localities by the following winter.

So is that redwing feeding on your pyracantha berries the same visitor as last year? Definitely not.

△ In common with other thrushes, redwings feed mainly on berries in the autumn and early winter.

Keeping low

When a predator such as a sparrowhawk strikes a flock of small birds, they often scatter. Redwings, however, often use a completely different method of escape – they simply stay on the ground where they are feeding and freeze.

It is thought that, because redwings frequently feed in leaf-litter on the ground, they benefit simply by squatting down. Then the sparrowhawk, or other predator, finds it difficult to distinguish these brown birds from the leaves around them.

Black redstart

Species: *Phoenicurus ochruros*
Family: Muscicapidae

◁ *The handsome black redstart is very much a bird of roofs, walls, buildings and industrial landscapes.*

IDENTIFICATION This is a black or brownish bird (14 cm/5½ in, the size of a great tit) that would be dull but for its orange tail, which is constantly shivered – a habit that is unique and is probably a type of signalling to others of its kind. The black redstart is quite an upright species, usually seen perched on roofs, walls or rocks.

The *male* is distinctive, with neat slate-grey plumage that becomes blacker towards the face, a bold white wing-bar and the signature orange tail.

The *female* is all over quite an intense sooty-brown, but for the tail and undertail. The wings are brown, without the wing-bar.

The *juvenile* (early to late summer) is browner than the female and subtly spotted on the breast; it moults to first winter plumage in autumn to look similar to the female. The following spring, first-year males resemble adults, but lack the white wing patch.

SHAPE AND CHARACTER This is an upright small bird with a large head and eye. It is easily identified by its habit of shivering its orange tail. It is not normally seen perched in trees, but on the ground and on buildings. Sometimes it runs along the ground, pausing frequently to scan for food. It is quite nervous, often flying from one spot to another.

VOICE Its main call is a loud 'tsip', often followed by 'tucc, tucc'. Its song is loud enough to be heard above the roar of traffic. It has two parts: an introductory twitter followed, a couple of seconds later, by a very odd, super-fast jangle of notes sounding like ball-bearings being rubbed together.

HABITAT This is really a mountain species that has adapted to nesting on buildings. In some areas, such as Britain, it is found principally in urban and industrial areas, but is a common garden bird in parts of continental Europe.

FOOD It eats invertebrates, including flies, spiders, ants and worms. It feeds by running and hopping over the ground, spying moving insects, or by watching from a raised perch and dropping down on whatever is detected. It will make short aerial flights after insects and occasionally digs down up to 4 cm (1½ in) for them. It takes a few berries in the autumn.

IN THE GARDEN Despite nesting in buildings, the black redstart tends to be indifferent to feeding stations or boxes.

BREEDING Its loose cup-nest is usually placed in a hole or crevice, as much as 45 m (148 ft) above ground. The female incubates four to six eggs for 12–16 days and both parents feed the chicks in the nest for 12–19 days. The young sometimes leave before they are able to fly, but can rely on being fed for at least 11 days after they fledge. There are often two broods, sometimes three.

MIGRATION It is resident in many parts of central and southern Europe, but a summer visitor to the east and to parts of Britain. It leaves breeding areas in mid-autumn and returns in early to mid-spring.

ABUNDANCE This bird is rare in Britain, more common in continental Europe.

BLACK REDSTART

Singing in the autumn

Male black redstarts sing in the spring, as do most other small birds, but also give a short burst of song in the autumn as well, which is unusual. The reprise occurs between the annual moult in late summer and the departure to winter quarters in mid-autumn.

This isn't just singing for singing's sake; it is territorial and has a mate-attraction function. But it is all done with an eye on the future. A black redstart singing in autumn is setting down a marker for the breeding season to come next year. As it sings it often acquires female company, and it is likely that the same female will pair with the singer the following year.

You might think that the autumn singing season would be an ideal time for a young bird of the year, a few months old, to make exactly such a mark. Strangely, this isn't the case. Only birds with at least one breeding season behind them take part.

High notes

The black redstart has a habit of selecting very high song posts from which to sing. These are often at the very top of buildings several storeys high, or atop machinery such as cranes, 50 m (165 ft) or more above ground.

These birds are also indefatigable singers. At the start of the breeding season they may sing 5,000 times a day – sometimes at night, too, by the light of street lamps or the moon.

Co-operation

Birds often throw up unusual cases of males and females co-operating in unusual ways. One good one concerns a male black redstart paired to two females. The two females both laid eggs (presumably), and they often alternated bouts of incubation, although at times they sat on the nest at the same time, side by side. When the eggs hatched one female brooded the chicks while the other helped the male to collect food. Later on, all three birds co-operated in feeding the chicks as they grew. It must have been tense, but it worked.

▽ The black redstart spends much time feeding on the ground, where its characteristic quivering of the tail distinguishes it from a robin.

Blackcap

Species: *Sylvia atricapilla*
Family: Sylviidae

IDENTIFICATION This is a distinctive small, beady-eyed, dark-brown bird (14 cm/5½ in, the size of a great tit) with an obvious grey tint to its plumage. It would be hard to identify, if not for its bold-coloured cap.

The *male* has a diagnostic black skull-cap.

The *female* has a toffee-brown skull-cap.

The *juvenile* (early summer to early autumn) is like the female, but young males may have a mixture of black and brown on the cap.

SHAPE AND CHARACTER The blackcap is rather slender, furtive and often difficult to see, in low cover or up in tree crowns. It is a very spirited and often aggressive bird, and spends time hopping from perch to perch within bushes. In common with other warblers, it is not sociable.

VOICE Its most common call is a sharp 'tack' or 'tack, tack'. Its song (early spring to mid-summer) is a pleasing, cheerful whistled phrase that characteristically starts with hesitant, under-the-breath mutters and ends powerfully, with a ditty of strong, clear notes.

◁ *The female blackcap is easily distinguished from the male (opposite) by its toffee-brown crown.*

HABITAT The blackcap is a bird of tall trees and shrubbery, found only in well-vegetated gardens or those with woodland nearby.

FOOD It feeds chiefly on insects in the summer and will take these at any time. It also feeds on fruit in autumn and winter, coming for berries such as honeysuckle and holly. It picks much of its food from leaves, but can also feed on the ground and low down.

IN THE GARDEN It will come to bird tables for a variety of foods, including crumbs and scraps, rolled oats, fat and grated cheese. It also visits berry-bearing bushes such as holly, honeysuckle and cotoneaster, and comes for apples.

BREEDING The blackcap builds a neatly constructed cup-nest, hidden in a low thicket or small tree, sometimes close to the ground. The clutch of four to six eggs is incubated by both sexes for 11–12 days. The young birds gain their feathers by 10–14 days, but are fed for at least two weeks after leaving the nest.

MIGRATION This bird presents a complicated migration picture. Essentially it is a 'leapfrog' migrant, with northern populations (including British ones) being summer visitors, wintering in sub-Saharan Africa. Many central European birds hardly travel at all, remaining close to their breeding grounds, so they are effectively resident.

ABUNDANCE It is fairly common.

BLACKCAP CONFIDENTIAL

▷ *Blackcaps are regular visitors to feeding stations in some parts of Europe, where they are notably aggressive to other birds.*

Changing migration patterns

Research on blackcaps has shown that migration is a more flexible behaviour than previously imagined. When a change in migration is favoured by natural selection, it can occur.

Indeed, laboratory experiments have shown that blackcaps have inheritable migratory tendencies. Some from central Europe migrate south-west to winter in the Mediterranean, while others go south-east. When such birds are cross-bred with each other, their offspring show intermediate tendencies, ending up going south.

Over the last 30 years or so blackcaps have been found wintering in England in increasing numbers. Rather than the individuals concerned being summer migrants staying on, their origin was traced to the breeding population in Belgium, Germany and Switzerland. Clearly such birds had flown west in autumn, instead of south-west, as most central European birds do.

But why change their migration route? Well, gardens probably enable blackcaps to survive better in the winter than they used to. And a blackcap wintering in England can cut 1,500 km (930 miles) off its migration route and arrive at its breeding grounds earlier in the spring.

Clues in the song

In some species an extraordinary amount of information can be conveyed in a male's song – in the case of the blackcap, the clue is not in the repertoire or vehemence of the delivery, but in the song rate.

In experiments in Austria, scientists found that a male blackcap's song rate early in the breeding season was somehow related to the subsequent fate of the nest. Males with high song rates suffered significantly lower rates of nest loss by predators later on. How could this be?

It turned out that, by singing at certain rates, male blackcaps were advertising the quality of their territory – in this case, the density of its vegetation. Naturally, the denser the vegetation, the more likely it was that predators would not find the nest. So it seems that male blackcaps can 'describe' their territory by singing at a certain rate. Unfortunately, there is a downside. It turns out that the most vigorous-singing blackcaps are also the laziest, contributing markedly less to incubation and chick-rearing.

Chiffchaff

Species: *Phylloscopus collybita*
Family: Sylviidae

◁ *Chiffchaffs are indefatigable singers, continuing their efforts well into the summer after most songbirds have fallen silent.*

IDENTIFICATION The chiffchaff (10–11 cm/4–4⅓ in, slightly smaller than a blue tit) is a small, plain-coloured olive-green bird that sometimes turns up in gardens. It is very active, constantly flitting about. Often it wags its tail with a downward flick. It has a thin, spiky dark bill and dark legs. The plumage is featureless, olive-green above and paler below, the most conspicuous feature being the pale eyebrow. The ear-coverts are quite dark, offsetting a pale eye-ring.

The *male* and *female* look alike.

The *juvenile* (early summer to early autumn) is browner above and yellower below than the adults.

SHAPE AND CHARACTER This is a very small, quite rotund bird with a spiky bill and round head. Typically for a warbler, it is very active, never remaining still for very long, and is good at concealing itself in foliage, although not as low down as some warblers. Its habit of repeatedly flicking its tail downwards is very useful. It is conspicuous when singing, selecting treetop song posts, often on dead branches.

VOICE Its most frequently heard call is a loud, far-carrying 'hweet!', with a slight upward inflection. Its famous song, giving rise to its English name, is best transcribed as 'chiff-chaff, chiff-chaff, chiff-chaff…' repeated 3–20 times in a series, at a steady pace. Between these series it often gives a few hiccupping 'terric' notes.

HABITAT It inhabits a species of broad-leaved woodland that can be found in larger gardens and may pass through smaller ones on migration.

FOOD The chiffchaff eats almost entirely small insects, including caterpillars, aphids and flies.

IN THE GARDEN It may come to a bird table for small scraps, especially in winter.

BREEDING It builds quite an intricate nest, low down or on the ground, often among dead leaves, but also in grass at times. There are five to six eggs to a clutch, which are incubated for 13–15 days by the female only. Once hatched, the young remain in the nest for 12–15 days. There are sometimes two broods.

MIGRATION It is typically a summer visitor, arriving in early spring and leaving late, by mid-autumn. The migration seasons of early to mid-spring and late summer to mid-autumn are when chiffchaffs are most likely to turn up in gardens. It is resident in parts of continental Europe, especially the south.

ABUNDANCE This bird is fairly common.

CHIFFCHAFF CONFIDENTIAL

Bedfellows?

Male and female chiffchaffs don't form a very close pair-bond in the woods where they breed. The female tends not to need much help in feeding the young, so the two seldom meet once the eggs have been laid. The male spends most of its time high in the canopy singing and feeding; meanwhile the female attends the nest based low to the ground and hunts for food down in the undergrowth.

Raising after raising

For a bird that spends so much time singing in the canopy, the ground seems a strange place for the chiffchaff to build its nest. However, in one study, 21 out of 55 nests were placed there. However, the ground has its perils, and studies in Russia show that the nests are usually placed high where mammal predators are common. And perhaps the birds learn the error of their ways – in the same study quoted above, only 4 out of 25 nests built for a second brood were on the ground.

Harbinger of early spring

The chiffchaff is one of a number of summer visitors that have shown a marked change in their migratory patterns in recent years. Most of these travellers show the same change: earlier arrival times and later departure times. The longer stay in Europe seems to be an indicator of global climate change. Why else would they stay longer if the temperature wasn't warming up?

In the case of the chiffchaff, the change can be measured in reference to mean first arrival date (as recorded at bird observatories). In the decade 1970–79 the mean first arrival date at British stations was 21 March; in the following decade it went back to 17 March; and in the seven years between 1990 and 1996 it was 11 March. Perhaps in the future chiffchaffs will come back earlier and earlier, until there is no point in them going in the first place?

▽ *A female chiffchaff approaches the nest. It leads a secretive life in the undergrowth, in contrast to the male's high profile.*

Goldcrest

Species: *Regulus regulus*
Family: Regulidae

IDENTIFICATION More diminutive even than a wren, the goldcrest (9 cm/3½ in) is Europe's smallest bird. It is often harder to see than identify: a minute, round-bodied bird with generally olive-green plumage, a black staring eye, a colourful crown-stripe (often difficult to see) and two pale wing-bars.

The *male* has some orange on the crown-stripe (most obvious when the crest is raised).

The *female* has a yellow crown-stripe bordered with black on either side.

The *juvenile* (early to late summer) has no crown-stripe, and its bill is paler.

SHAPE AND CHARACTER The goldcrest has a distinctive shape, with a rounded body, large head and short tail. It never stops moving, flitting rapidly from perch to perch; it gives a busy impression by constantly flicking its wings. It also frequently hovers in front of branches like an overweight hummingbird. It is often surprisingly tame, allowing close approach.

VOICE It has a very high-pitched 'see' call with a hissing effect – a common sound of autumn. The goldcrest sings an equally needle-sharp cyclical phrase in the breeding season, which sometimes incorporates imitated fragments of other bird sounds.

HABITAT At heart this is a bird of coniferous woodland, but a single large tree can sometimes support a breeding pair. Outside the breeding season it is often found in deciduous trees, and even bushes, in gardens.

FOOD The goldcrest specializes in very small insects such as springtails, aphids and caterpillars, plus small spiders. It may feed constantly throughout the day, working the branches.

IN THE GARDEN Occasionally the goldcrest visits fat-balls and bird tables for crumbs. Plant conifers for the nest site.

BREEDING It begins breeding later than many other small birds, but may still raise two broods. It builds a remarkable dense cup-nest, suspended from a tree fork. The female lays six to eight eggs, which are incubated for 16–19 days, considerably longer than for other small birds. The young leave the nest after another 17–18 days and may be cared for by the male while the female builds a new nest.

MIGRATION It is mainly resident. Breeding birds from Scandinavia migrate south, and some arrive in mid-autumn to winter in Britain.

ABUNDANCE The goldcrest is fairly common where there are conifers.

◁ *Goldcrests are acrobatic foragers, often clinging upside-down from foliage and frequently hovering. They also regularly creep up bark, like treecreepers.*

GOLDCREST

Master builders

Goldcrests build intricate and long-lasting nests. Although only used in one breeding season, the nests frequently survive longer than their builders, remaining woven into hanging coniferous twigs for several years. They are therefore more than suitable for bringing up small young in the cold northern parts of Europe.

The nest is a cup of three separate layers, two of moss and an inner one of insulating materials such as feathers and plant down. Remarkably, it is so warm and well camouflaged that goldcrests have higher breeding success than their larger peers in the northern conifer forests. Even heavy snowfall and violent storms cannot thwart these minute birds. The nest is also high flexible – the outer layer is bound together with cobwebs – meaning that it won't buckle to the elements, or to the strain of the growing young birds. When a goldcrest feeds its brood, nestlings that are satisfied retreat to the bowels of the nest and stay there, reducing the heat loss that would occur if all nestlings remained at the top and kept the nest entrance open.

Tough bird

Anybody who has ever seen a goldcrest feeding hungrily on a winter morning with snow on the ground cannot help but admire the resilience of Europe's smallest bird. It has been known to survive even when the outside temperature drops to -25°C (-13°F) and to breed up to the Arctic Circle. Even so, its resilience comes at a price. In a harsh winter, mortality is very high. Individuals have to spend up to 95 per cent of the daylight hours feeding constantly, and at night several birds will roost together in bodily contact. On a cold night they may still lose 20 per cent of their body weight.

Migratory feats of endurance

This midget's migratory behaviour is also astonishing. Many northern birds migrate south in autumn, moving at night. Total journeys of up to 2,400 km (1,500 miles) have been recorded (for example, Estonia to France), and they can fly more than 240 km (150 miles) in a single go. Goldcrests look so fragile that these feats are hard to believe.

▷ The yellow, black-bordered crown above a plain face readily identifies a goldcrest. This is probably a female, since males usually show a dash of orange on the crown.

Pied flycatcher

Species: *Ficedula hypoleuca*
Family: Muscicapidae

IDENTIFICATION The pied flycatcher (13 cm/5 in, slightly smaller than a great tit) is a diminutive, neat and active songbird. It is distinctive for the strong contrast between its dark upper parts and clean white/whitish underparts and for its large white wing-bar. It is usually seen making darting, fly-catching sallies into the air and back again. It is quite large-headed, with a dark, staring eye; long wings and quite a short tail with very narrow white sides. It has black legs.

The *male* is very black-and-white: a clean jet-black on all of the upper parts except for a small spot on the forehead; clean white underneath, with a large white patch on the wings.

The *female* is essentially brown where the male is black; it lacks the white forehead patch, and is a little dusky on the breast, with a smaller wing-bar.

The *juvenile* (early to mid-summer) is, apart from its shape and obvious wing-bar (tinged brown), very different from the adult, with small spots above and below.

The *first winter* is, by late July, similar to the female.

SHAPE AND CHARACTER The pied flycatcher is a compact, but long-winged sprite with a restless, agile manner. Its darting action is typical. When perched, it often flits its wings and cocks its tail.

VOICE Its main call is a sharp 'whit', with a hint of goldfinch call about it. The male's song (mid-spring to early summer) is a curious ditty of a few rather strained notes, delivered with confidence until the end. The song sounds a little like another bird song played backwards.

HABITAT The pied flycatcher lives in deciduous and mixed woodland, spilling over into orchards, parks and gardens, so long as there isn't any dense human settlement adjacent.

FOOD As you would expect from the name, the pied flycatcher feeds on flying insects, in particular flies, bees, wasps, ants and beetles, plus caterpillars and, in late summer and autumn, a few berries. It uses a sallying-darting method, often returning to a different perch after its foray. It also collects food directly from leaves, or even from the ground.

IN THE GARDEN It takes so readily to nest-boxes that it actually prefers them to natural sites. Use an enclosed box with up to a 5 cm (2 in) entrance hole. It needs a perch near the nest-box.

BREEDING In the wild the pied flycatcher uses a hole in a tree for nesting. The female incubates six to seven eggs for 13–15 days and once hatched, they are fed by both parents, but mainly by the female. They leave the nest at 14–17 days. Each female has just one brood, but males are often polygamous.

MIGRATION It is a migrant to tropical West Africa, arriving to breed in Europe in mid- to late spring and leaving by late summer to early autumn. Many European birds stop off in northern Spain and Portugal before feeding up and leaving Europe.

ABUNDANCE This is an uncommon bird.

△ *A female pied flycatcher is much more soberly patterned than the male, with smaller white wing markings.*

▷ *The handsome pied flycatcher is a shy, restless songbird that is difficult to see in the tree canopy – despite the male's bold plumage coloration.*

PIED FLYCATCHER

◁ *The brownish plumage coupled with a large white wing-bar suggests that this singing bird is a male pied flycatcher without the usual colour scheme.*

Cuckoldry

There is some natural justice in the tale of deception by the male (opposite). Whenever a male moves away more than 10 m (33 ft) from its primary female, it immediately introduces a considerable risk that another male will take the opportunity of its absence for a spot of extra-pair copulation. Thus only about 75 per cent of the primary brood is likely to be fathered by it.

Mate parade

The pied flycatcher is one of the few species in which it is known how many partners an individual bird may 'try out' before selecting a particular one for pairing. In this species it is the female that chooses, by visiting singing males on their territory.

It is fewer than you might expect – just four on average. The problem is that, in a competitive environment, pairing needs to be speedy. A female doesn't wish to waste too much time searching when other females might be sizing up the same possibilities.

Colour changes

Male pied flycatchers are not always black and white. Some are dark brown and white instead, looking quite similar to the females.

In a few parts of Europe the pied flycatcher and the closely similar collared flycatcher overlap in range. Where that happens, the male pied flycatchers sport the brown plumage so that females are able to distinguish them from male collared flycatchers.

Stopovers

After the breeding season, pied flycatchers migrate to West Africa for the winter. But they don't go there straight away. Instead, the whole European population stops over for a few days in northern Spain and Portugal, to fatten up. Once they have almost doubled their fat-free weight, they complete the rest of their journey, over the Mediterranean and the Sahara, in a single flight.

Deception in the male

In recent years it has become routine for biologists to uncover cases of 'infidelity' in birds, where a male or female will routinely copulate with individuals other than its social mate. It is also reasonably common for individual birds to be bigamous – for a male to sire two broods with different females, for example. However, while the pied flycatcher is commonly bigamous (the rate varies from 2 to 39 per cent of the population), there is a difference. In the case of this smart black-and-white bird, the bigamy comes with an unusual degree of deception.

Pied flycatchers pair up in the usual way, with a male singing within its territory to attract a female. Once this is done, the female lays eggs and incubates them for a couple of weeks. During this time the male is often absent. It turns out that, most unusually, it is not just pairing up with another female – it is doing so in a second territory in another part of the wood. This territory is, on average, 200 m (650 ft) away from the first territory, but may be as much as 3.5 km (2 miles) away.

The reason for the distance is that it wants to deceive the secondary female into thinking it is unpaired. A female would probably be unwilling to pair up with a male it knows to be already 'attached' – and with good reason. Usually a male abandons its secondary mate entirely and the abandoned female suffers from poor breeding success as a result. So the deceiving male uses the distance from its first territory to increase the likelihood that the second female will not find out about its previous activities.

△ *A male pied flycatcher approaches the nest with food. Not all females are fortunate enough to receive this help.*

▽ *An intimate portrait of a pair of pied flycatchers belies the true nature of many relationships in this species.*

Spotted flycatcher

Species: ***Muscicapa striata***
Family: Muscicapidae

◁ *The spotted flycatcher is well known for catching large insects, including bees, butterflies and, in this case, a moth.*

IDENTIFICATION This small brown bird (14 cm/ 5½ in, the size of a house sparrow) is easiest to identify by its behaviour, periodically darting out from a perch to catch an insect in flight. It has long wings and a long tail, making it look streamlined. The bill is long and wide and the legs are short. It is rather dull brown above, pale below, with obvious dark streaks on the throat and forehead. The wing feathers are pale-edged.

The *male* and *female* look alike.

The *juvenile* (mid- to late summer) has pale spots on the back and rump.

SHAPE AND CHARACTER The spotted flycatcher is streamlined, with long wings. Its mouth is wide to enable it to reach and catch insects in flight. Its characteristic feeding method involves watching – relatively motionless – from a perch, then flying out to snap a passing insect and returning to the same or another perch. It often perches on dead branches high in the tree canopy. Alternatively, it may hunt from a low perch such as a garden fence.

VOICE Its main call is a quiet 'wiss-chook'. It also makes a high-pitched sound, like someone breathing with a blocked nose. Its song is a repetition of several astonishingly squeaky and scratchy notes, well spaced in a series; it is easy to miss, and not pleasing to listen to.

HABITAT This is a bird of woodland glades, which finds flower-rich gardens highly suitable.

FOOD It eats mainly flying insects, especially blowflies and hoverflies; it also takes butterflies. In poor weather it may feed on the ground.

IN THE GARDEN The spotted flycatcher won't use a feeder, but is a good example of a bird that benefits from pesticide-free, flowery borders. Unlike the pied flycatcher, it often uses open-fronted nest-boxes or a nest shelf.

BREEDING It is a late breeder, but often fits two broods into a summer. The nest is a loose cup of twigs, grasses and roots on some kind of sheltered ledge, often among creepers. The female lays four to five eggs and incubates them for 13–15 days; the young hatch after another 13–16. Both sexes feed the young in the nest and for up to a month afterwards.

MIGRATION The spotted flycatcher is migrant to tropical Africa. It is one of the latest summer visitors to arrive, rarely before late spring, and the last birds are usually seen in early autumn.

ABUNDANCE The spotted flycatcher is fairly common, but widely declining.

SPOTTED FLYCATCHER

CONFIDENTIAL

Fly-catching

Careful studies have shown that the spotted flycatcher adapts its insect-catching according to the conditions and time of day. In the early morning it tends to feed high in the tree canopy, on swarms of tiny insects such as midges or aphids, which it will often glean from the leaves. Another important morning food is bumblebees; these insects are among the few available early on, but have the disadvantage of being difficult to handle. Before eating them, the birds have to beat them on a perch to remove the sting, a process than can take up to a minute. Although bumblebees and wasps make a good meal, this handling time makes them hardly worth the while and, when the day warms up, flycatchers readily shift to blowflies and other, more harmless insects. When it is warm they almost invariably use their method of flying out from a perch.

On rainy days, when insect supplies tumble, spotted flycatchers switch to feeding on the ground, flying down onto insects that they spot or even just hopping about. They are not very good at this and suffer badly in poor weather. They are really fair-weather birds and tend to seek out the early-morning or late-evening sun at the beginning or end of the day.

Stay-at-homes

For a bird that is a long-distance migrant, breeding in Europe and wintering in sub-Saharan Africa, the spotted flycatcher certainly seems to like familiarity. It has been shown that, both on its breeding and wintering grounds, it comes back to the same places again and again. Observations in Africa have found the same individuals returning to wintering areas for at least three consecutive years, and remaining in a small area throughout.

This so-called 'site fidelity' is even more pronounced in the breeding areas. In Europe a series of different individuals have been known to use exactly the same breeding territory for 48 consecutive years.

Short stages

Britain's two common flycatchers have interestingly different migration strategies. The pied flycatcher spends time stocking up on fuel at stopovers to make progressively longer migratory flights as it goes. The spotted flycatcher, however, progresses in shorter stages, accumulating only as much fuel at each stopover as it needs to reach the next. Even in the Sahara, it depends on finding food for its journey at convenient oases.

▷ *Spotted flycatchers often build their nests among creepers on the walls of buildings. The garden environment suits them well.*

Long-tailed tit

Species: *Aegithalos caudatus*
Family: Aegithalidae

IDENTIFICATION This is a tiny-bodied but very distinctive long-tailed bird (14 cm/5½ in, of which the tail is 7–9 cm/2¾–3½ in). It has an unusual colour combination of black, white and pink, a very small, short bill and a red eye-ring.

The *male* and *female* are similar. On the head, a black eyebrow broadens towards the nape and joins the mantle, leaving the rest of the head white.

The *juvenile* (late spring to mid-summer) is similar to the adult, but most of the side of the face is dark brown, with only the throat and crown being white.

SHAPE AND CHARACTER Its diminutive size, long tail and habit of feeding within trees and bushes make for a unique combination. The long-tailed tit is rarely seen alone, but moves through vegetation in flocks of 5–20 that remain together all day long. It is rather tame. Other birds often join up with long-tailed tit flocks.

VOICE The long-tailed tit does not have a loud territorial song, but just makes soft contact calls of three kinds: a short 'tupp', a louder splutter and persistent 'see-see-see-see' calls.

HABITAT This is really a woodland bird, but visits gardens freely. It tends to be seen in the mid-levels of trees.

FOOD It specializes on tiny items, even as small as the eggs and pupae of butterflies and moths, plus spiders, flies and bugs. In winter it takes a few seeds.

IN THE GARDEN Flocks of long-tailed tits sometimes visit fat-balls and hanging nut-feeders.

BREEDING It may begin construction of its intricate nest in early spring, and the process takes three weeks. The large nest is domed, with a side-entrance, and is placed low down in a (thorny) bush or high in a tree fork. There are six to eight eggs, incubated for 15–18 days, and the young leave 16–17 days after hatching.

MIGRATION The long-tailed tit is resident, making local movements only.

ABUNDANCE It is common in more rural gardens and in those close to woods.

◁ Once a rare visitor to feeding stations, long-tailed tits have learned to visit them in some parts of Europe over the last 15 years.

▷ Truly unique, the long-tailed tit cannot be mistaken for anything else. Note the minute bill, which is used for gleaning the smallest invertebrates from stems and leaves.

LONG-TAILED TIT

CONFIDENTIAL

◁ *A long-tailed tit brings a feather towards its nest. It will probably need another 1,000 of these.*

Death brings life

The long-tailed tit's need for feathers to line its nest is acute in the building season: to get adequate insulation a pair needs 1,000–2,000 of them in all. But where can a small bird obtain so many feathers? Certainly not from its own plumage – that would compromise its survival.

Although some birds are fortunate enough to build close to, for example, a roost site of larger birds such as geese or ducks, which provides feathers aplenty, the answer is usually, remarkably, a corpse. Life and death are everyday events in the garden, and it only takes one dead bird – especially a larger one such as a pigeon – to provide all of a long-tailed tit's needs. This is a superb example of one bird's demise becoming an opportunity for another.

Helpers at the nest

The effort involved in feeding young birds in the nest is proverbially heavy: witness the 1,000 visits to the nest a day undertaken by a pair of blue tits. Small birds take great risks in energy costs and potential predation during their two- to three-week sprint of child-feeding.

Imagine, then, the sheer advantage to be gained if more than two birds were to help at the nest. Normally this doesn't happen, because helpers reap no benefit. But among long-tailed tits – birds with an unusual social structure – extra help is a frequent event. Between one and nine helpers have been recorded at a single nest, mucking in with the parents to feed the young. Not surprisingly, such extra help invariably increases the number of young successfully fledged.

But why do long-tailed tits help at others' nests, when the personal benefits of doing so are zero? The answer stems at least partly from kinship. Helper birds are almost always related to the male pair member. Typically they have failed to breed successfully themselves early in the season, their nest having been destroyed or plundered. For these birds, failure can be mitigated by helping out at the nest of a brother.

◁ *Most small birds avoid any kind of bodily contact because of the risk of disease and feather lice transmission. However, long-tailed tits huddle to roost together at night.*

Huddles

Long-tailed tits are not closely related to the other tits and one of the things they do very differently is to roost communally. Typically, tits such as blue and great tits go to sleep on their own in a hole, except for females on the nest with young.

Among long-tailed tits, communal roosting is a survival tactic. On cold winter nights the birds huddle together in bodily contact and thus cut down on their personal heat loss. When roosting, the birds gather on a horizontal branch in a thick bush, with the dominant birds in the middle and the subordinate ones towards the outside. Despite the fluffy appearance of the group, the strict hierarchy means that the birds on the outside are most likely to be the first to die of cold.

The birds roosting together at night are based on the winter flock – a pair of adults with their young of the previous year. There are often other adults too, usually relatives of the male member of the founding pair. Their presence is significant: they may be the very same birds that helped raise their brood the previous breeding season. If so, the enhanced survival opportunities obtained by being a member of a flock might well be their 'reward' for hard labour the previous summer.

◁ *In order to insulate its nest properly, the long-tailed tit needs to find a lot of feathers. Where better than on the corpse of a bird?*

Great tit

Species: ***Parus major***
Family: Paridae

◁ *The bold and colourful plumage of the great tit, with white cheek surrounded by black, makes it readily identifiable.*

▷ *Great tits are among the commonest and most adaptable of all garden birds, often the first to find a newly put up feeding station.*

IDENTIFICATION Boldly patterned, this impressively colourful garden bird (14 cm/5½ in, the size of a house sparrow) has a black stripe down its yellow breast and prominent white cheeks surrounded by a black and grass-green back. Its bluish wings have a white wing-bar, and its blue tail has white outer tail feathers.

The *male* has a black stripe down the front that is broader and neater than the female's and goes right down to the vent. The broadness of this stripe is an indicator of social status: females prefer males with bigger stripes.

The *female* has a narrower black stripe, which peters out on the belly.

The *juvenile* (early to late summer) is a duller version of the adult, with a yellowish cheek.

SHAPE AND CHARACTER The great tit is unremarkable in shape, but bigger than other tits. It is bold and active, and often aggressive at feeders and elsewhere. It feeds in tit-like fashion in trees, hanging upside-down at times. It is also inclined to feed on the ground. Its flight is stronger and smoother than the weak flitting of the rest of its family.

VOICE It makes a bewildering range of calls, from scolds to a chaffinch-like 'pink-pink'. From mid-winter to early summer it produces a loud, cheerful and forceful 'teacher, teacher, teacher' song, with great variation in pitch and tempo. This is a dominant sound in garden neighbourhoods in late winter.

HABITAT This is a woodland bird that has adapted well to the garden environment.

FOOD The great tit is essentially insectivorous (especially caterpillars) in the breeding season and herbivorous (seeds and nuts) from autumn to late spring.

IN THE GARDEN The great tit is a regular visitor to all types of feeding stations, attracted by nuts and seeds, plus meat bones and fat. It routinely uses enclosed nest-boxes with a 2.8 cm (1 in)-plus entrance hole and a depth of at least 12.7 cm (5 in) – a standard tit-box.

BREEDING It manages one brood a year, with egg-laying in mid-spring. The adults select a tree-hole or other cavity and the female builds a cup-nest. Seven to nine eggs are incubated by the female, which is fed by the male, for 13–15 days, and the young leave the nest 18–21 days after hatching. They are independent after a couple of weeks.

MIGRATION This bird is resident. The young disperse up to a few kilometres away after fledging. In northern Europe large movements occur when the population is high.

ABUNDANCE It is very common in most gardens.

Changing its tune

Great tits in urban areas have been shown to have higher-pitched songs than their rural counterparts, because these can be heard more easily above the din of traffic, enabling the birds to communicate more effectively. Surprisingly, this characteristic is so entrenched that urban birds respond sluggishly to the songs of rural individuals played over a loudspeaker, while reacting strongly to their neighbours.

Strange foodstuffs

Great tits are not without their predatory side. They have occasionally been observed killing small birds – indeed, one was seen to kill a goldcrest and carry it off in its 'talons'. They also eat large insects and sometimes small lizards or frogs.

Recently an unexpected trait was discovered in a cave in Hungary. Great tits were seen seeking out and eating roosting bats. Similar behaviour has been suspected near bat roosts in Poland and Sweden. Pipistrelle bats, the main victims, are only a quarter of a great tit's size.

◁ *No wonder this male great tit looks worse for wear; when feeding the young it might have to find and deliver 500 caterpillars a day.*

▷ *Although great tits feed mainly on insects and nuts, they are not averse to the odd foray into predatory behaviour. Birds, bats and lizards have all found themselves on the menu.*

△ *While large body size is generally an advantage for an individual bird, it has been found to be a problem for making a quick escape from predators. So dominant great tits are relatively small.*

Dawn chorus

Great tits are enthusiastic members of the dawn chorus – that short bout of vigorous singing at first light on spring days, which involves many species at once. Research into the intensity of their singing has cast new light on the function of the dawn chorus, which for a long time has puzzled biologists. Researchers found that the most intense dawn singing in male great tits involves a specific stage in the breeding cycle, when the females are laying the clutch of eggs. This also coincides with the female's most fertile period, when there is the greatest risk that it might pursue extra-pair copulations with a neighbour; at this time the male has to be particularly alert to this danger to its paternity and guards the female carefully.

This suggests an intriguing possibility. Perhaps the chorus itself, in great tits at least, proves to the female that its mate is awake, active and ready for copulation. The facts seem to bear this out. Once the dawn chorus stops, the female frequently comes out of the nest-hole and copulates – presumably with whichever bird is closest to hand.

The perfect meal

Although great tits work hard bringing caterpillars to their large brood of developing young, they also hunt intelligently. It is not just a case of bringing the first caterpillar you find. Instead, the foodstuffs are sorted. Newborn chicks are only fed on tiny caterpillars, while the adults eat any larger ones. As the chicks get older, the caterpillars they are given get bigger, until the sixth day when size no longer matters. But this is not a hard-and-fast rule. It depends on how far the parents have to travel to get food. If they find themselves making longer commuting journeys, they will bring in larger caterpillars than they would otherwise.

Top dogs are trim

Great tits live in an unequal world where certain individuals are dominant over others. Intuitively, one would expect that dominance would be measurable in terms of physical attributes, especially body size. And, in certain situations, the dominant birds are indeed larger and heavier than their peers. However, this fails to take into account the effects of predation. For whereas heavier birds may be more effective in skirmishes over food, trimmer birds are better at evading sparrowhawks. As a result, in populations where sparrowhawks are common, dominant great tits can actually be smaller than their rivals. This isn't a problem, though. Being socially dominant means that they still have access to food whenever they need it.

Blue tit

Species: *Cyanistes caeruleus*
Family: Paridae

IDENTIFICATION This is a very small, abundant songbird (11.5 cm/4½ in), yellow below, with a narrow dark stripe on the lower belly. Its face pattern is bold, with a thin stripe through the eye, a cobalt-blue cap and white cheeks surrounded by a blue collar. Its wings and tail are blue, and its back greenish.

The *male* and *female* are alike.

The *juvenile* (early to late summer) is like the adult, but duller and with yellowish cheeks.

SHAPE AND CHARACTER This small, round-headed, dumpy bird is irrepressibly agile and lively, noisy and aggressive. It is also bold and one of the most frequent users of bird feeders. It often hangs upside-down when feeding (its legs are very strong), and tends to move quickly everywhere. Its flight is weak and flitting. It is sociable and often joins flocks of small birds.

VOICE The blue tit makes a bewildering range of calls and songs. Its calls include a distinctive multi-syllable scold and a gentle 'si-si-si'. Its main spring song consists of a drawn-out three-note introduction followed by a lively, silvery trill.

HABITAT This is essentially a bird of broad-leaved woodland that has successfully adapted to gardens. The blue tit often feeds at the tops of trees, but will also feed in hedges and bushes.

△ *The cobalt-blue cap is a key identification of a blue tit. The blue has a strong ultraviolet component that the birds use to choose a mate.*

▷ *One of the most abundant of all garden birds, the blue tit habitually visits feeding stations in large numbers.*

FOOD It feeds mainly on caterpillars in the breeding season and gives them to its young. In the autumn and winter it takes seeds and nuts from the ground and a few berries from bushes. Its strong feet enable it to be acrobatic in the treetops and obtain food from leaves and stems.

IN THE GARDEN The blue tit is a regular visitor to all types of feeding stations, attracted by nuts and seeds, plus meat bones, fat and scraps – it is often the very first bird to find a new source. It routinely uses enclosed nest-boxes with an entrance of 2.5–3.5 cm (1–1⅓ in).

BREEDING In the breeding season it usually attempts just a single brood, in contrast to most small birds, which may have two or more. It tends to lay a large clutch of 10–12 eggs (and up to 18) in mid-spring. Fed by the male, the female incubates for 13–15 days. As soon as the eggs hatch, the parents collect astonishing numbers of caterpillars – up to 1,000 a day – for their young, which leave the nest after 16–22 days and are independent shortly thereafter.

MIGRATION This bird is essentially sedentary.

ABUNDANCE It is very common in most gardens.

If the cap fits

For humans, all blue tits look pretty much the same – that's because we cannot see in the ultraviolet (UV) spectrum. In this realm the brightness of various parts of the plumage differs significantly from bird to bird and can play a part in the sexual attractiveness of an individual.

This applies particularly to the cap. Males have brighter caps than females, and some individuals have much brighter caps than others. Laboratory experiments have shown that females have a preference for brighter-capped males, so it seems this feature plays a part in mate choice. This is perhaps not surprising, because the cap is often raised in social encounters and the head tilted forward.

Further experiments revealed that both sexes seemed to select for a certain reflectance. Within pairings, the two individuals tended to have the same ratio of UV reflectance to normal light reflectance – so-called 'UV-purity'. This is cap compatibility, you might say.

The secret of good sons

An extraordinary finding from recent research is that blue tit females can skew the sex ratios of their broods. If they have particularly good-looking mates, they will conceive more males. The advantage then passes to the grandchildren: in terms of overall breeding success, it is better to produce high-quality males than high-quality females, because the males will, in future, pass on the genes by copulating with other females outside the pair-bond. In a Swedish study, the 'sexiest' males (based on their UV reflectance) had broods that were 70 per cent male, while poor-quality males sired chicks that were 70 per cent female.

◁ *The blue tit is one of the most intensively studied birds in the world. Apart from its abundance, it also has a habit of using nest-boxes and this helps researchers to trap and ring birds.*

△ *The breeding season of a blue tit is very accurately timed to coincide with a glut of caterpillars in the late spring woodland.*

The secret of a good father

Another aspect of a blue tit's plumage provides a clue as to how good a father it will be – the intensity of yellow on its breast.

Yellow is a pigment that is quite difficult to synthesize. It comes from the diet, and scientists have shown that, in blue tits, the individuals with the brightest yellow breasts are most successful at catching and eating caterpillars. Could this favourable trait be an indicator of how good a certain male is in providing for its chicks?

To check this, scientists from Barcelona carried out an ingenious experiment. They took 70 clutches of eggs from known parents and switched them to the care of foster parents in which the yellowness of the male's breast varied. At the age of two weeks they measured the growth rate of each chick. They found that the largest and best-fed chicks came from nests in which the foster fathers had the most intense yellow breast, while there was no correlation with the breast colour of the genetic father. Thus it was proven that intensely yellow-breasted birds were indeed the best providers.

Processed food

Have you ever wondered why, when feeding their young, tits only bring in one caterpillar at a time? Why don't they – like blackbirds – gather a bundle of wriggling bodies and feed the mass to more than one chick at a visit?

The reason seems to be the need to process the food first. Caterpillars are essentially jaws on legs, and the larger ones are potentially strong enough to damage the mouth of a tit nestling. Thus the parent tit hammers a potentially hazardous caterpillar with its bill to take its head off, before delivering it to a chick. Parent tits will also, on occasion, remove the gut of the caterpillar in the same way – thereby probably discarding tannins in the caterpillar's innards, which are known to reduce the growth rate of tit nestlings.

Coal tit

Species: *Periparus ater*
Family: Paridae

IDENTIFICATION The coal tit (11.5 cm/4½ in) is the size of a blue tit, but not brightly coloured, just a mixture of black, white and brown. Look out for the white 'badger-stripe' nape patch, the stained brown underparts and two narrow but noticeable white wing-bars made up of spots.

The *male* and *female* look alike.

The *juvenile* (early to late summer) has a yellow tinge to the cheeks and nape.

SHAPE AND CHARACTER This bird is very small, with a large head and a spiky tail; note the very thin bill. It is typically active and agile, frequently feeds upside-down and has a habit of momentarily hovering in front of branches, like the goldcrest. It has a tendency to feed high up in tree crowns. It is often harassed by other tits, so makes only brief visits to feeders.

VOICE Its main call is a surprisingly loud 'dwee' with an upward inflexion. It has a similar song to the great tit – a repetition of two notes in a series, like the sound made by a foot-pump. The coal tit's version has less attack on the first note and sounds markedly less cheerful.

HABITAT This is a bird of coniferous woodlands and is mainly found in conifers in the breeding season. Outside this time it may travel more freely and turns up in gardens that only have deciduous trees.

FOOD In the breeding season it eats mainly caterpillars and other insects. In autumn and winter it takes seeds, especially of spruce, but others too. It tends to feed high up in the branches, minutely examining the needles and cones.

IN THE GARDEN It is a regular visitor to bird feeders, ranging from trays to hanging nut-dispensers, for peanuts, seeds and fat. Try putting a standard nest-box low down on a conifer tree.

BREEDING The coal tit breeds in a hole in a tree, a wall or even down by tree roots, building a cup of moss and spiders' webs. Unlike other tits, it regularly raises two broods of nine to ten eggs. They are incubated for 14–16 days and the young leave the nest 16–19 days after hatching.

MIGRATION It is mainly sedentary, but some continental birds move south and west in the autumn.

ABUNDANCE The coal tit is a common bird.

◁ *A small, fairly soberly coloured tit with a large head and short tail, the coal tit can be distinguished by its two white wing-bars.*

COAL TIT

Food storing

Coal tits tend to be quick and furtive when they visit a bird table or hanging feeder, dashing in, grabbing a nut and dashing away again. They live the life of a subordinate, being at the bottom of the tit pecking order, trampled by great tits, blue tits, crested tits, marsh tits and willow tits. When a coal tit is part of a mixed feeding flock, it can expect a fellow member to steal food from it every three minutes or so.

But the coal tit does have a secret weapon against this abuse. It stores food away from the eyes of its competitors, in quiet corners of its territory, in lichen or bark crevices, tree holes or on the ground. If there is plenty of food around, it will immediately spirit some away for consumption later on. It takes a risk by doing so, for by acting alone it no longer benefits from the many eyes and ears of the flock. But the risk of predation is outweighed by the benefit of being able to feed alone in peace.

In this form of food storing, the coal tit differs from the willow tit, which stores food for the coming winter months. The coal tit invariably retrieves its food stores within 48 hours.

Hanging around

The tit family is well known for its ability to hang upside-down, practising the art not only on hanging bird feeders, but also up in the trees. It's a useful way of reaching hard-to-get-at places, such as the underside of twigs and hanging leaves, which are out of reach to larger birds. The morphology of the hind leg is specially adapted to aid flexion in habitual hangers, such as the coal tit.

However, hanging is still quite an energy-consuming activity, as is shown up in individual differences between coal tits. Larger, fatter individuals hang upside-down less often than smaller individuals, and birds tend to hang less frequently at the end of the day, when they have built up fat reserves to carry them through the night.

◁ *Coal tits never linger at feeders for long, but usually dash quickly away. If they tarry they might be attacked by other tits.*

△ *A coal tit stores a nut away in a mossy bank. It will retrieve it any time between a few minutes or a couple of days later.*

Marsh tit

Species: *Poecile palustris*
Family: Paridae

IDENTIFICATION Despite its English name, this is a smart woodland tit (11.5 cm/4½ in, the size of a blue tit) with what would be quite distinctive plumage if it weren't for the remarkably similar willow tit. It is essentially brown above, soft buff-brown below, with a bold, glossy black cap, black moustache and white cheeks. The bill is black except for a white mark on the inner part of the upper mandible (the jaws that make up the bill), at the cutting edge.

The *male* and *female* look alike.

The *juvenile* (late spring to late summer) looks similar, but with a more sooty cap and paler below.

SHAPE AND CHARACTER This is a small, well-proportioned songbird with a short, fairly broad bill. It is similarly acrobatic to other tits, well able to hang upside-down from branches and search anywhere in leaves. It tends to search lower down in trees than the blue tit or coal tit, in the middle layers. It often holds a nut or seed in its feet on the perch and hammers at it with its bill. It is quite an aggressive species that is dominant over the coal tit and willow tit.

VOICE It has a diagnostic call that helps to distinguish it from the willow tit: a sharp, slightly sneezing 'pit-chou!' Its song is like a clipped version of a great tit's song, without the cheery feel: 'chip, a-chip, a-chip, a-chip…'

HABITAT The marsh tit is found in broad-leaved woodland, especially with a healthy understorey, and so tends to be found in gardens adjoining this habitat.

FOOD It feeds on insects in the breeding season, but for much of the year subsists on a vegetarian diet, including large numbers of nuts and seeds, plus a few berries. In many areas beech mast is very important. The marsh tit stores food.

IN THE GARDEN It comes readily to bird tables and hanging feeders for seeds and nuts. It will sometimes use a standard tit nest-box.

BREEDING The marsh tit's nest site is a natural hole in a tree or tree stump. The female builds a shallow cup of moss, in which it lays a clutch of six to ten eggs, white speckled with reddish-brown. The incubation period is 13–15 days, typical for a small bird, and the young leave the nest 18–20 days after hatching. There is usually just one brood a year.

△ The smart, well turned-out marsh tit lacks the wing-bars of the coal tit. Note the pale mark at the cutting edge of the upper mandible, a key characteristic separating it from the willow tit.

MIGRATION It is very much a non-migrant, with most birds remaining in their territory all their lives.

ABUNDANCE Locally the marsh tit is common.

MARSH TIT

Stay-at-homes

Have you ever seen a flock of marsh tits? If you think so, perhaps you should look again – it is unusual to see this species in any numbers at all.

The marsh tit has moved far away from the familiar image of the sociable tit. It does things differently. Rather than being sociable and mobile in autumn and winter, it lives in pairs in a fixed territory all year round. Indeed, once a pair have a territory, they are unlikely to leave it for the rest of their lives; they also pair up for life in a truly monogamous relationship.

This isn't to say that marsh tits don't join tit flocks – they do. But they only join them when the flock passes through their own territorial borders. Once the flock reaches the edge, the pair retreat back to their own ground.

Clutch control

The marsh tit exhibits a number of traits in egg-laying and incubation that are typical of tits as a whole. For example, it tends only to lay one large clutch of eggs a season, timed precisely to coincide with the bloom in caterpillar numbers in its native woodland.

The female lays one egg a day, but does not usually begin incubation until the last egg is laid. What controls the exact number of eggs is very subtle, and has been the subject of much research. It seems that the female has some way of predicting how many young it and the male will be able to bring up in any given year, and lays that many eggs.

Only the female incubates them, while the male brings in regular portions of food. However, that doesn't stop the incubating bird having breaks. It usually sits for 50 minutes or so every hour, then takes a break for ten minutes before returning to duty, having fed a little and defecated.

The first and last eggs to be laid are the most vulnerable to problems and often don't hatch. For reasons unknown, the eggs containing male chicks are also less likely to hatch that those containing females.

▽ An adult marsh tit approaches the nest with food. Prior to starting their clutch, marsh tits seem to 'know' how many chicks they can raise in a season.

Willow tit

Species: ***Poecile montanus***
Family: Paridae

◁ *Willow tits have a slightly larger head and 'muscular' neck than marsh tits, and most show a pale panel on the wing.*

▷ *The willow tit is an exceptionally hardy species – it can even dig tunnels in the snow in which to sleep.*

IDENTIFICATION This is a neatly plumaged tit (11.5 cm/4½ in, the size of a blue tit) that is exceedingly similar to the marsh tit. Like that species it is brown above, soft buff-brown below, with a bold, glossy black cap, black moustache and white cheeks. Most individuals show a pale wing-panel caused by pale edges to the secondaries. The bill is all black.

The *male* and *female* are alike.

The *juvenile* (late spring to late summer) is similar, but has a more sooty cap and is paler below.

SHAPE AND CHARACTER This is a typical tit with quite a large head, a short bill and strong legs for feeding acrobatically on trees and plants. It is rather a quiet, unobtrusive species that often, in contrast to other tits, doesn't call for long periods. It tends not to hack at nuts in the manner of the marsh tit.

VOICE This can be useful in distinguishing it from the marsh tit. It lacks the latter's explosive call, instead giving a drawn-out, rather buzzing 'eez-eez-eez'. Its song (mid-winter to mid-spring) is also distinctive, a simple repetition of pure notes: 'pu-pu-pu...' with a downward inflection. It is infrequent compared to that of other tits.

HABITAT It is found in several discrete habitats, including coniferous woodland (where the marsh tit doesn't occur), damp, marshy wooded areas with willow, alder and birch, and hedgerows. Gardens on the edge of any of these habitats could have this species.

FOOD In common with other tits, it feeds on insects in summer and seeds and other plant material in autumn and winter. With its weaker bill it takes some smaller seeds than the marsh tit.

IN THE GARDEN It tends to visit bird tables or feeders only sparingly, for nuts. It rarely uses standard-sized tit-boxes, which must be filled with sawdust or other loose material so that the birds can 'excavate' it. Place them low on a birch tree, or try importing a rotting birch/alder stump to the garden.

BREEDING The willow tit actually excavates its own nest-hole for a breeding attempt, rather than using an existing hole. It hardly builds much of a nest at all; the absence of moss distinguishes it from other tit nests. The usual clutch is six to eight eggs, which are incubated for 14 days. The young remain in the nest for a further 17 days, tended by both parents.

MIGRATION The willow tit is not migratory. Occasionally large numbers of birds move down from the far north of Europe, however.

ABUNDANCE This is an uncommon bird.

WILLOW TIT

Northern nights

The willow tit is an exceptionally tough species, being one of the few small birds to survive winters right up above the Arctic Circle. This bird might look small and weedy, but it clearly isn't.

The techniques that it uses to survive the winter nights are nothing short of remarkable. First, the willow tit is unusual among birds for its ability to drop its metabolic rate at night. This falls fully 5°C (9°F) below standard body temperature and naturally enables the bird to use up less internal fuel than it would otherwise have to.

Even more extraordinary, perhaps, are the roosting sites that willow tits use. At the beginning of winter they almost always use unoccupied rodent burrows, usually in the ground. But as winter becomes harsher, they switch to making holes in the snow. They can dig a 20 cm (8 in) tunnel in just 10–15 seconds, but the speed belies the hole's usefulness. Adults usually occupy the same tunnel night after night, and even throughout the winter. They can break snow to enter it at the end of the day and can similarly dig themselves out of fresh snowfalls in the morning.

Cold storage

Another remarkable aspect of the willow tit's ecology is its habit of storing away food items. These stored items are retrieved in times of food shortage, which naturally usually occur during the winter.

The sheer scale of food-hoarding defies belief. In the winter, Russian birds can store 200 items per day, with many more in the spring. It is estimated that, over the course of the year, a single bird will spirit away nearly half a million seeds and nuts among clumps of lichen, in crevices in the bark and on the ground. Some remain in storage for months on end.

Perhaps even more remarkable is the bird's memory. It has been shown that it retrieves 95 per cent of the items it stores. That is a feat that even the most gifted human beings would do well to match.

Crested tit

Species: *Lophophanes cristatus*
Family: Paridae

IDENTIFICATION The crested tit (11.5 cm/4½ in, the size of a blue tit) is completely recognizable because of its unique tuft-like crest. Its plumage is mainly dark brown above, soft buff-brown below. Its head is boldly patterned with a black throat, neck ring and a C-shape around the ear-coverts (cheeks). The crest itself is scaly black and white, blacker in summer.

The *male* and *female* look alike.

The *juvenile* (late spring to late summer) is similar, but lacks the neck collar and has a much shorter, blunted crest.

SHAPE AND CHARACTER This would look like an average small bird if it weren't for the crest. It is quiet and unobtrusive, so easy to miss. It is not sociable, but is seen in pairs or family parties. It is as acrobatic in the branches of trees. Its flight is weak and flitting; it doesn't go far.

VOICE Its main call is an understated, gentle, slightly bubbling trill, easily missed, but very characteristic. It also makes quiet 'zit' calls. Its song is a combination of the two: 'zi-zit, prr-rr-rr'.

HABITAT It is found primarily in coniferous woodland (pine in Scotland).

FOOD As with other tits, it eats both animal and plant food: insects (including caterpillars) and spiders, and seeds (chiefly from conifers). It feeds a great deal from the surfaces of trunks and large branches, often adopting a treecreeper-like foraging style, and also forages frequently on the ground. It caches large amounts of food.

IN THE GARDEN The crested tit will come sparingly to bird feeders for nuts and seeds. It can sometimes be persuaded to use a nest-box with an entrance hole of 2.9–3.8 cm (1¼–1½ in).

BREEDING The crested tit makes a cup-nest of moss in a hole in a tree stump or tree, usually in rotting wood, often low down. It lays three to nine eggs, which are incubated for 13–16 days by the female, provisioned by the male. The young leave the nest after 18–22 days. Usually there is just one brood.

MIGRATION It doesn't migrate. Adults remain in their territory for all their lives, while the young disperse short distances in late summer.

ABUNDANCE Locally the crested tit is common.

◁ *No other small bird in Europe has a tuft-like crest like this, so the crested tit is unmistakable.*

▷ *Crested tits are most closely associated with coniferous woodland, where they spend much time quietly searching the surface of branches and trunks for hidden insects.*

CRESTED TIT CONFIDENTIAL

Heading back

The pattern on the back of the crested tit's head – a brown nape with a black border that forms a point at the base of the raised crest – happens to be almost identical to what you see when looking at a crested tit from the front. This 'false face' is thought to confuse predators.

Tenants

A study in Sweden has uncovered an interesting social arrangement among crested tits in the winter. The population seems to consist of small groups of two, three or four birds, which hold and defend a collective territory. Anything above two seems to consist of an adult pair, with either one or two first-year birds, which – perhaps surprisingly – are not related to the adults. This system brings to mind human households with live-in lodgers.

If there are two first-year birds in the flock, they always seem to be a male and a female, and could be thought of as a pre-breeding pair. However, another fascinating statistic suggests that a different outcome is frequent. Apparently, 89 per cent of all first-year birds get their initial breeding opportunity not by pairing with one of their peers, but with an adult whose mate has died.

▽ In contrast to most other tits, crested tits are very quiet and only give their distinctive trilling call sparingly.

◁ *Crested tits habitually store food away in hiding places within their territory. This means that, if food is short at some time during the winter, they can retrieve what they have put away.*

Headless corpses

It's a curious fact that, in Europe, all the widespread brown-plumaged tits (coal, crested, marsh and willow) store food, while the colourful species (blue and great) do not.

In the case of the crested tit, the food is stored away for both long-term and short-term use. A bird generally stores food as soon as it finds it, taking it away to be hidden in the bark of a tree, always in a unique hiding place. It will then be retrieved when food is otherwise scarce.

You might think that a tit would only store away imperishable items such as seeds and nuts, but this isn't the case. Crested tits sometimes store insects as well, including caterpillars; most are killed by pecking, and their bodies may be 'glued' inside a bark fissure with the bird's saliva, its own body fluids or even by binding it with cobwebs. Some caterpillars may only be paralysed by the act of decapitation. Such unfortunate larvae may remain alive and fresh for a week or more before finally being eaten.

▷ *The pattern on the back of a crested tit's head is almost the same as that of a bird facing forwards. This could confuse predators.*

Nuthatch

Species: *Sitta europaea*
Family: Sittidae

IDENTIFICATION This is a distinctive woodland bird (14 cm/5½ in, the size of a great tit) that creeps close to branches. It is blue-grey above, pale buff below, darker towards the vent, with white spots. It has a bold black mask through the eye. The tail has white corners.

The *male* has bright-chestnut flanks to its vent, obviously contrasting with the belly.

The *female* has a less intense chestnut colour, intergrading to the belly.

The *juvenile* (early to late summer) has browner upper parts, duller flanks and its mask is not as bold.

SHAPE AND CHARACTER This is a woodpecker-like small bird, with a large head, short tail and a long, strong pointed bill. However, its tail is not used as a prop; instead, the huge feet enable the nuthatch to hold on tight to tree trunks and branches and move up and down. It is the only bird that climbs down trunks head-first. A restless, cheery bird, it moves quickly around a tree in a jerky fashion. It is often aggressive at bird feeders. It flies with a strong undulating flight, looking like a woodpecker. It can feed on the ground.

VOICE Its most common call is a cheery, human-like whistle 'hwit-hwit-hwit …', which can be slow or fast-paced. Its various songs include a slow 'pew, pew, pew' with a mournful ring, but also a super-fast trilling.

HABITAT Gardens with nuthatches require a surrounding habitat of tall, mature deciduous trees, usually close to a wood.

FOOD It takes insects and spiders obtained by searching the surfaces of trunks and the thicker branches of trees. In autumn and winter it eats large quantities of seeds such as hazel, acorn, beech and pine. It often wedges a seed into a bark fissure and hacks it open with its bill.

IN THE GARDEN The nuthatch is a frequent visitor to hanging bird feeders and feeding trays, and is fond of nuts and seeds. It will often use a standard tit-box with an entrance hole of 2.9–3.8 cm (1¼–1½ in).

BREEDING Its nest sites are holes in trees, or sometimes in walls. As a preliminary stage, the female gathers mud and plasters the nest entrance until it is the right size for a nuthatch, but nothing larger. Six to eight eggs are incubated by the female, which is fed by the male, for 16–18 days, and the young leave the nest 24–25 days after hatching.

MIGRATION It is very sedentary. Once settled in a territory, pairs may not leave for the rest of their lives. The young disperse in mid-summer.

ABUNDANCE Locally the nuthatch is common.

◁ *Notice the very large, strong feet on this nuthatch. These are an adaptation to a life of climbing trees.*

▷ *Nuthatches have no hesitation in using feeding stations in gardens. They are often aggressive to other visitors.*

Exacting requirements

Nuthatches are so fussy about the materials they require for nesting that they remind one of a well-heeled human couple intent on constructing their perfect home. The nuthatch's nesting needs must certainly limit where these birds are able to keep a territory.

The first requirement is for large, deciduous trees in which there are suitable holes. Second, the builders need a supply of mud in their woodland habitat so that they can modify the hole entrance and make it an exact fit. Third, for reasons that are unknown, nuthatches like to line their nests with the bark of coniferous trees, usually pines. No other garden birds need such an eclectic combination of building materials.

The reason for the plastering is almost certainly twofold. First, the birds modify both the entrance hole and the interior of the cavity to an exact size that fits only them. Smaller birds that might be competitors are easily repelled, while larger competitors, such as starlings, are excluded by the size of the entrance. The other reason is predator deterrence: the mud plaster is copious and hard, and it would take a determined woodpecker to chisel the plasterwork away and predate the eggs and young.

Looking down

Nuthatches are the only bird family in the world able to forage by climbing head-first down the trunks of trees. Plenty of birds, including treecreepers and woodpeckers – and nuthatches themselves – can climb upwards to forage, but the rest would overbalance if they tried to climb down. The nuthatch's specialized clinging technique, holding one foot above the other and using the upper one as a pivot and the lower one as a support, enables it to pass off this trick. Having a short tail that is not used as a prop gives it the flexibility to do so.

But why climb downwards in the first place? Perhaps looking downwards gives the nuthatch a different perspective and a competitive advantage in seeing insects lodged within the cracks and fissures of tree bark?

◁ The nest of a nuthatch is in a hole, the entrance of which is usually plastered up with mud to keep larger birds out.

▷ Nuthatches are unique among birds for their ability to creep down tree limbs head-first. However, this doesn't stop them creeping upwards and sideways, too, if the fancy takes them.

Treecreeper / short-toed treecreeper

Species: *Certhia familiaris / C. brachydactyla*
Family: Certhiidae

IDENTIFICATION The treecreeper (12 cm/4¾ in) is the size of a small tit, but has an elongated body shape and distinctive behaviour. It is always seen unobtrusively creeping up (but not down) trunks and branches, hugging the bark. It is richly spotted and speckled in various hues of brown above, contrastingly pearly-white below. It has a distinctive white supercilium (eyebrow). In flight, the wings appear strongly barred.

The *male* and *female* look alike.

The *juvenile* (early to late summer) is similar but distinctly colder-toned in plumage.

SHAPE AND CHARACTER The treecreeper has a very distinctive slimline shape, with a long, curved bill and a long tail that acts as a prop when the bird is climbing. It spends most of its time climbing up tree trunks and branches, hugging them closely and climbing with jerky hops. It frequently and abruptly switches tree trunks by flying down from one to the base of an adjacent one, with a weak-looking, undulating flight.

VOICE Both species makes high-pitched calls in a distinct series. The short-toed treecreeper's call is fairly loud; while the treecreeper's is easily missed. Their songs (mid-winter to early summer) are different: the treecreeper's is a soft descending phrase with upward inflection at the end. The short-toed treecreeper's song is a shorter, more cheerful phrase and doesn't descend much.

△ The short-toed treecreeper is only subtly different from its Eurasian cousin (shown opposite). It has a duskier brown belly, clearer white tips to the primaries, a slightly longer bill and shorter hind claw.

HABITAT The treecreeper rarely strays far from wooded habitats, so tends to favour large gardens or those close to woodland. It uses any kind of tree – deciduous or coniferous – for feeding.

FOOD The treecreeper takes a variety of insects and spiders at all times of year, with a few small seeds (mainly spruce and pine) in winter. Its main feeding method is simply to forage on the trunk surface by creeping upwards (never head-first down like the nuthatch) and searching crevices and fissures. It often ascends in a spiral.

IN THE GARDEN It can be attracted to fat and suet, especially if smeared onto tree bark. Very occasionally it uses tit-boxes for a nest. It is possible to obtain specially designed nest-boxes for this species, but they don't seem to work very often.

BREEDING The treecreeper builds a nest on a tree trunk or within ivy. It lays a clutch of five to six eggs, incubated by the female for 13–15 days. The young are tended by both parents for 14–16 days in the nest and for some time thereafter when they have left it. When threatened, the young 'freeze', head up against the bark, and are very difficult to see.

MIGRATION This bird is a non-migrant in most areas.

ABUNDANCE It is fairly common in large gardens or those close to woodland.

TREECREEPER

Keeping its tail up

The tail of a treecreeper is an important part of its climbing apparatus, being used as a prop when the bird is in the vertical position. The tail feathers are specially strengthened so that they can support it against gravity.

But what happens during the moult? All feathers need to be changed at least once a year. In most small birds, the tail feathers are moulted gradually from the middle outwards – the same feather on each side in sequence. However, in treecreepers this would compromise their climbing, so the central feathers remain in place until all the other feathers have been changed.

Keeping production up

The breeding season is often a rush, and treecreepers demonstrate this as well as any garden bird. In common with many smaller birds, they attempt to cram two broods into a single year. But this is difficult and often leads to something of a compromise.

The first brood is fed in the nest for a total of 14–16 days. However, it is only for the first ten or so that the female is a provider. After this it gives up, to concentrate on finishing a new nest that the male has already started, while the male takes over food deliveries to the chicks. When the first brood finally leaves the nest, the female will be out of their lives and they rely on the male until they are independent. Even as they make their first flight, the female could be incubating the second clutch.

Keep off

Recent studies have shown that, although treecreepers will climb up any type of tree in their territory, there are some that they keep away from – those above a nest of wood ants. It isn't that they are avoiding these aggressive insects, but that the ants tend to remove most of the insects the birds would feed on. It is a rare example of measured competition between an insect and a bird.

△ *Treecreepers forage by climbing up tree limbs. Their white underparts are thought to reflect light into the bark fissures where invertebrates may be lurking.*

Magpie

Species: *Pica pica*
Family: Corvidae

IDENTIFICATION The magpie (44–46 cm/17–18 in, roughly pigeon-sized) has a long tail and iridescent glossy wings and tail, with purplish and greenish hues. The tail is longer in the middle (graduated).

The *male* and *female* look alike.

The *juvenile* (late spring to mid-summer) is duller, less iridescent, with a shorter tail.

SHAPE AND CHARACTER This bird is distinctive, with its long tail that is flirted and flicked upwards constantly. It is quite bold and approachable; it flies off with unsteady flight on rapid wing-beats, seemingly making poor progress. It tends to be seen in pairs, but small flocks make up a significant part of the population. It often just sits in a tree, moving its tail, but also spends much time on the ground.

VOICE It is noisy, making a loud, scolding 'cha-cha-cha', sometimes in machine-gun-like bursts. It makes a variety of quieter conversational calls.

HABITAT It is found in lightly wooded country and more open areas with scattered trees for nesting. The garden environment is ideal for it. In places it occurs in more urban areas, apparently helped by regular kills on the roads.

FOOD An omnivore, the magpie feeds on a variety of animal and plant material: in spring and summer mainly on insects (especially beetles), found by wandering over the ground, but also taking the eggs and nestlings of birds and a few small mammals. It takes much waste material – even dog faeces – and often feeds on corpses. Fruits and seeds make up the vegetarian part of its diet.

IN THE GARDEN The magpie occasionally visits the bird table for bread and scraps, sometimes fruit. It nests in garden trees. It is often unpopular among garden enthusiasts for its habit of taking the eggs and young of other birds, but its overall effect on garden bird numbers is negligible.

BREEDING The characteristic stick nest, domed in structure, is a common sight in suburbia. The single brood of five to eight eggs is laid in mid-spring and incubated for 21–22 days. The young fledge after 22–28 further days and once out of the nest, are looked after for about six weeks.

MIGRATION Basically the magpie is non-migratory, although there may be a few local movements.

ABUNDANCE It is a common bird.

◁ *Magpies are often seen in flocks, which are made up from birds that don't have a territory.*

▷ *Contrary to the opinion of those that think of magpies as killers, these birds spend most of their time foraging on the ground for invertebrates.*

MAGPIE

◁ *Recent studies have shown that magpies are able to recognize their image reflected in a mirror.*

Ceremonial gatherings

Magpie society has two strata: there are pairs that hold territories and live and breed in them; and then there are the rest, including young birds, single birds and pairs that have lost a home – all of which may spend time together in flocks with a loose association, although they cannot breed.

This arrangement doesn't make life easy. It divides into the 'haves' and 'have-nots', and every so often the 'have-nots' try to strike a blow against their landed rivals. Their uprising is public and ritualized and is known as a 'ceremonial gathering'. It begins with a trespass. Dominant males within a non-breeding flock crash through the borders of a territory and begin to display and posture. Their commotion soon attracts spectators, both the non-breeders and nearby pairs, which may number up to 20 individuals. It also attracts the incumbent pair and, as the parties square up, there is a great deal of noise and chasing and flirting of tails.

In the event, nothing much happens. Usually the interlopers are rapidly evicted, although just occasionally this act of subversion works and a new pair gains a foothold within the original pair's territory. If this happens, it can be expanded little by little, until the two pairs become neighbours.

Reflections

Recent studies have shown that magpies can recognize themselves in a mirror. Red and yellow stickers were placed on captive birds' plumage in places where the magpies could not normally see them. The birds were completely unaffected by this treatment until they saw their own reflection; only then did they attempt to remove the stickers, using their feet and bill.

Stolen reputation

Magpies have a reputation for stealing shiny objects and hiding them away, to the horror of human beings who have left out jewellery such as wedding rings. The habit is even commemorated in an opera by Rossini.

Apparently this is entirely a myth. They just don't do it, at least not in the wild. They do store things away, but only edible items.

Arch enemies

The magpie has a fractious and violent relationship with the carrion crow. It could be said that the enmity is caused by the two species' own similarity. They are closely related, and compete with one another for food and nest sites.

It tends to be the magpie that comes off worst. Whenever the two species are feeding close by, carrion crows chase and harass magpies and reduce the length of time that they can feed. In one study, magpies attended feeding sites for an average of 2.1 minutes per session when no carrion crows were around, but this reduced to 1.7 minutes when carrion crows were present at the same time.

During breeding, carrion crows are one of the major nest predators of magpies, so that magpies nesting close to crows have very little chance of breeding successfully. In rural areas, magpies often nest close to houses, because the presence of humans puts the wary crows off.

Construction industry

The large stick nest of the magpie is often a common sight along suburban streets. It lasts through the breeding season and into the following winter, appearing well built to withstand a battering by the weather.

Despite this durability, magpies usually build a new nest construction every year. It is quite a complicated task, involving moving large numbers of twigs or sticks (600 or more) and a great deal of mud. At its best its nest will be a large, strong dome with a 15 cm (6 in) wide cup in the middle.

However, not all magpie nests are the same, depending on their history. If built by a pair that live permanently on a territory, construction can begin in the first mild days of January and continue for 40 days or more, a little at a time. Some nests, however, made by newly formed pairs may be put up in a week. Often such rushed nests have their roof missing!

Egg transfer

Several species of garden birds lay eggs in the nests of other birds. While everyone has heard of the cuckoo, which lays eggs in the nests of different species, more often a bird will lay extra eggs in the nests of neighbours of its own species.

Magpies in America sometimes do this, but with an extraordinary twist. Rather than laying *in situ*, these birds apparently lay the egg in their own nest and then transfer it in the bill to their neighbour's family.

▽ Magpies build very solid nests and take a great deal of trouble constructing them – or at least, they usually do.

Jay

Species: *Garrulus glandarius*
Family: Corvidae

◁ *The stunning jay, with its four-colour plumage, is completely unmistakable.*

▷ *Jays are famously shy birds that regularly visit gardens in the early morning when most humans are asleep. They make a very loud, harsh, alarm call.*

IDENTIFICATION The jay (34–35 cm/13½–14 in, roughly the size of a feral pigeon) is an unmistakable, colourful, but shy bird of woodlands and gardens. The jay has four colours: much of its body is greyish-pink, but it has a black tail, a white rump, a black moustache and black-and-white wings. The crown is streaked. The corner of the wings has an extraordinary patch of sky-blue barred black.

The *male* and *female* look alike.

The *juvenile* (late spring to mid-summer) has smaller crown streaks and is a little redder.

SHAPE AND CHARACTER The jay has a typical crow shape, with a large, straight, multi-purpose bill. Its flight is characteristic, fluttering with fluent wing-beats that give it a weak, stuttering progress; it swoops down to the ground. It is a shy bird of the treetops that nonetheless feeds a great deal on the ground. It is easily alerted and flushed, and is usually seen in pairs.

VOICE It has a noisy, discordant screech that alerts the whole wood or garden to danger. When several jays discover a predator such as a goshawk or owl, the noise can be almost deafening. It also makes imitations, such as the call of the buzzard, and can even mutter soft notes under its breath.

HABITAT This is a woodland bird that also occurs in parks and large gardens.

FOOD The jay eats a wide variety of plant and animal food. In spring it takes large numbers of insects, especially caterpillars and beetles, and will also raid birds' nests for eggs and nestlings. Occasionally it catches and eats small mammals. In autumn and winter its diet is dominated by nuts, especially acorns, which it collects in large numbers.

IN THE GARDEN Jays will visit bird tables, often early in the morning, for vegetable scraps. Some will come to hanging nut-feeders.

BREEDING It makes a cup-nest out of twigs, placed in a tree fork. The female incubates four to five eggs for 18 days and the young are fed in the nest by both sexes until they fledge 21–22 days later. Once they have left the nest, the young are tended for up to eight weeks. There is one brood a year.

MIGRATION The jay is mainly resident. A few northern birds move south in the autumn.

ABUNDANCE Locally this bird is common.

JAY CONFIDENTIAL

Acorn collectors

Jays are omnivorous, but every autumn they concentrate their foraging on just one particular food source: acorns. On autumn days small groups may spend the daylight hours collecting nothing else. Acorns, the fruits of oak trees, are tough and long-lasting, and make ideal items for storing away in case times get hard in the winter.

They make good use of the acorn season, each bird collecting around 3,000 of these fruits between early and late autumn and secreting them in its permanent territory. Not all birds have sufficient oak trees nearby to collect enough, so they may have to commute to a wood further afield to fulfil their needs, often becoming obvious as they flop their way back and forth. Some birds commute up to 4 km (2½ miles); they may carry as many as nine acorns at a time, in bill and oesophagus, although most only carry three.

Having collected so many acorns, it would be madness to store them all away in one place, in case a competitor such as a squirrel were to find the hoard and plunder it. Instead, each acorn or small group of acorns is hidden away in a different hiding place – buried in the ground, placed in a bark fissure, and so on. It seems that each jay knows its territory so well that it can refind most of the hiding places, apparently by remembering landmarks.

Extraordinarily, the jay may not need every acorn that it stores. In a mild winter it might not use its hoard much. In that case, the acorns stay in the ground and may well sprout an oak sapling – a future food source for subsequent generations of jays.

▽ *Jays are famous for their habit of collecting acorns or other nuts in the autumn and storing them away in hundreds of hiding places for later consumption.*

Jays and squirrels

The jay's loud, discordant, screeching alarm call is a familiar sound in the woods where this bird lives. It is well known that other bird species respond to the call, because it is an effective early warning of the presence of predators. The interaction has earned the jay the lyrical nickname of 'Guardian of the Forest'.

A recent study has shown that it is not just birds that respond to the jay's alarm call. Red squirrels do, too. In tests using playback of the call, about half the squirrels being observed reacted with an escape response.

Neighbour trouble

Jays are well known as predators of the eggs and nestlings of other birds. They get less blame than magpies, however, because their exploits tend to be secretive.

Back in the 1950s it was shown that jays also suffer from egg predation of their own. This happened particularly in dense breeding populations. The culprits? Other jays.

▷ *When in flight, the jay makes rather fluttery, swooping, unsteady progress, which is characteristic.*

◁ *Jays are frequent robbers of the nests and eggs of other birds, but they don't get the same bad press as magpies, probably because they are more furtive.*

Jackdaw

Species: *Corvus monedula*
Family: Corvidae

IDENTIFICATION The jackdaw (33–34 cm/13–13½ in, the size of a feral pigeon) is much the smallest of the black crows, and perhaps the most distinctive, with its grey nape ('shawl') contrasting with the dark crown, and its white staring eye. The bill is also shorter than that of other crows. It is intensely sociable.

The *male* and *female* look alike.

The *juvenile* (early to late summer) is browner-tinted, with dark eyes and hardly any grey on the head.

SHAPE AND CHARACTER This bird is obviously a crow, with its strong legs and thick bill, but is smaller than the others. It flies with faster wing-beats, with the pace of pigeon rather than being slow and regal. It is very sociable, often seen in large gatherings, in which pairs stick together. It has a jaunty gait on the ground. In the air, jackdaws often indulge in aerobatics. Birds roost communally, often calling and wheeling about until it is nearly dark.

VOICE The jackdaw's main call is a sharp, friendly 'chack!' with many variations.

HABITAT It requires a combination of grassland and fields for foraging, and holes in quantity for nesting communally. It often uses buildings for the latter (sometimes chimneys) and so can succeed well in rural, urban and suburban landscapes.

FOOD The jackdaw is omnivorous, taking a wide variety of plant and animal foods, the latter including many insects in the spring and summer, plus birds' eggs, small mammals and reptiles, carrion and scraps from rubbish. It feeds by wandering over the ground and picking up food.

IN THE GARDEN This bird is shy, but often visits bird tables for a variety of scraps, such as fat, bones, potato, cereals, berries and nuts. It frequently nests in chimneys, but will also use a large enclosed nest-box with an entrance hole of 15 cm (6 in).

BREEDING It brings up one clutch a year, laying eggs in mid- to late spring. In vertical holes, it simply throws down sticks until they lodge, often building up an accumulation that can be a fire hazard in chimneys. It lays four to six eggs, which are incubated by the female for about 20 days. Once they have hatched, the young are fed by both parents and fledge at 28–36 days.

MIGRATION The jackdaw is mainly resident and non-migratory. Some northern European and upland birds move south in early to mid-autumn, returning in late winter and early spring.

ABUNDANCE This is a common bird.

▽ For birds that customarily nest in large holes in trees, a chimney makes an excellent alternative for a jackdaw.

▷ Jackdaws pair for life and the bond between the sexes is extremely strong. They seem never to be 'unfaithful'.

'Smug marrieds'

It is routine to see jackdaws in pairs, and not just in the breeding season, either. Even above winter roosts, when hundreds of birds may be wheeling in the air, it is easy to make out how many birds remain in their units, flying close by side by side, apparently inseparable. Indeed, the pair-bond in the jackdaw is known to be lifelong.

But is it as intimate and close as appearances suggest? There are many cases in which birds ensconced in close pairings, such as house sparrows, nevertheless indulge in extra-pair copulation without batting an eyelid. Are jackdaws any different? A recent DNA fingerprinting study shows that they are. Despite the fact that jackdaws are semi-colonial, with many non-breeding birds about and plenty of opportunity for extra-pair liaisons, experimenters did not find a single case.

Hard work

One reason behind the jackdaw's strict monogamy could be the recent finding that it is exceedingly difficult for jackdaws to feed their young sufficient food to survive. Although both sexes bring in as much food as they can, it seems to be a general rule that many of the chicks starve in most seasons.

The slobs are on top

A study of a jackdaw colony published in 2004 revealed a quite extraordinary finding: the most socially dominant jackdaws – those that had the first pick of food and other resources – were the least fit males in terms of their reproductive output. Despite being evidently the strongest and fittest individuals, they consistently produced fewer fledglings than their peers, and those fledglings they did sire had a lower survival rate. Furthermore, their mates laid smaller eggs and were in poorer condition than the equivalents in subordinate pairings.

As yet there is no proven theory for this bizarre finding, although it is possible that the dominant jackdaws' elevated testosterone levels might have suppressed their care of its mate and young.

Rook

Species: *Corvus frugilegus*
Family: Corvidae

IDENTIFICATION This big black bird (44–46 cm/ 17⅓–18 in, the size of a crow) is often seen in large flocks on farmland and grassland, nesting in tree colonies. It is easy to mistake it for a crow, but is more sociable and has an obvious bare, greyish-white skin patch at the base of the bill. Its plumage is glossy black all over with a purplish sheen. It has a steep forehead, which, along with the pointed bill and bare skin, gives it a distinctive profile. The feathers on the belly are long, making the bird look as though it is wearing baggy shorts.

The *male* and *female* look alike.

The *juvenile* (late spring to mid-summer) is browner and lacks the pale patch of skin, so it is almost indistinguishable from a crow. Through autumn and winter it gradually acquires its greyish-white skin patch.

SHAPE AND CHARACTER The narrow bill and high forehead are unique among black crows. In flight the rook is graceful, with longer outer wings that are more obvious 'fingers' than the carrion crow's, and it flies with more fluency and slightly faster wing-beats, although the difference is very subtle. The rook is an extremely sociable crow, feeding, roosting and breeding communally. It sometimes undertakes aerobatics, especially just before roosting.

VOICE It makes a typically crow-like caw, but this lacks the angry edge of the carrion or hooded crow. There are many variations, and at the colony it may make high-pitched sounds, like a voice breaking.

HABITAT This was originally a steppe and grassland species that has adapted well to farmland. It requires a combination of fields for feeding and tall trees for nesting. Within the vicinity, it wanders into towns and gardens.

FOOD The rook, like other crows, is omnivorous. It mainly forages over the ground for invertebrates such as earthworms and plant matter such as cereal seeds, probing deeply with its bill. It also takes some carrion and scraps, but is not as predatory as the carrion or hooded crow.

IN THE GARDEN Try putting out hanging bones, fat and other meat; it also visits bird tables for scraps.

△ The dirty white bill of the rook instantly distinguishes it from the other black-coloured crows. Note the 'baggy shorts' as well.

▷ Rooks are extremely sociable at all times of the year. In the spring, they breed in treetop colonies known as 'rookeries'.

BREEDING The rook is well known for its treetop nesting colonies known as 'rookeries', which contain anywhere between a couple of pairs and several hundred. The main pairing season is in the autumn. Two to six eggs are often laid by early spring and are incubated for 16–18 days. The young are fed by both sexes on worms and leave the nest 30–36 days after hatching, continuing to beg on the fields for some time afterwards.

MIGRATION In many places (including Britain) the rook is resident, but it may also migrate south in mid- to late autumn, remaining within temperate central Europe. It returns in later winter to early spring.

ABUNDANCE This is a common bird.

Clever rooks

Rooks in captivity are adept tool-users and are exceptionally intelligent. In one experiment, birds loaded stones into a long glass tube so as to raise the water level and reach a floating worm – just as recounted in Aesop's fables of 2,000 years ago.

In another experiment rooks were able to bend a length of wire into a hook and use it to obtain a bucket laden with food at the bottom of a glass tube. These members of the crow family can even use two tools in succession. In one experiment rooks needed to drop a small stone down a tube to release a trapdoor and thereby obtain a tasty meal. However, they first had to obtain the small stone, and could only do this by dropping a larger stone down a wider tube to release the trapdoor that unleashed the smaller tool. Using two tools in succession is known as 'meta-tool use', and is only seen in the great apes (including ourselves) and members of the avian crow family. Amazingly, there don't seem to be any records of rook genius in the wild, so the experiments on captive birds were quite a surprise.

Rookery rape

Colonies of rooks, with their close-set nests, are short on privacy and long on temptation. Rookeries provide an environment where liaisons with the opposite sex outside the pair-bond can be frequent. No doubt many such encounters are intentional, but rookeries happen to be of interest because of the high incidence of what can only be described as rape.

Female rooks on the nest are quite frequently pestered. It is thought there is a reason for this: the normal incubation posture assumed by the birds is strikingly similar to their solicitation display – a slight droop of the wings and a lowering of the head. Clearly, some neighbours simply cannot help themselves, and attempts are made on each sitting female several times a day. The act often causes a furore, as neighbouring males swoop down and try to interrupt the deed. Yet sometimes, it seems, the well-meant intervention dissolves into what it seeks to prevent – another attempt at rape. As an end result, the genetic material in each rook clutch is anybody's guess.

Carrion crow

Species: *Corvus corone*
Family: Corvidae

IDENTIFICATION This is the familiar all-black crow (45–47 cm/17¾–18½ in) of lowland western and central Europe, including most of Britain. It has an unmistakably menacing appearance, with a large, stout bill, a flat crown and strong legs. Its plumage is entirely black, without much gloss; the eye is dark.

The *male* and *female* look alike.

The *juvenile* (early to late summer) is similar, but the plumage is browner.

SHAPE AND CHARACTER This is a well turned-out crow with neat plumage in contrast to the rook's scruffy look. Its legs are tightly feathered. It has a regal style of flight with slow, unhurried wing-beats. It soars less often than the rook (or raven); its wings are a little shorter. It is markedly less sociable than the rook, but flocks do occur. However, it is most often seen alone or in pairs (which would be unusual for a rook) and nests solitarily. It usually walks nimbly across the ground, but may skip like a vulture.

VOICE Its familiar angry, rasping 'caw' is often uttered in a series of three, really belted out. It also makes a variety of quieter sounds, especially between pair members.

HABITAT It inhabits open country with scattered trees and woods, and does well in gardens, parks and cities.

FOOD The carrion crow is omnivorous, but more predatory than the rook or jackdaw. It will catch small mammals and eat birds' eggs and nestlings, and occasionally it kills larger birds. This bird frequently eats carrion and is a regular visitor at rubbish dumps and abattoirs. However, it usually eats more invertebrates, such as worms and beetles, than anything else. The carrion crow also takes much vegetable matter, including seeds, berries and cereal grain.

IN THE GARDEN The carrion crow comes to the bird table and ground feeding stations for a range of scraps, including meat, fat and potatoes. It often dunks bread in the pond before eating it.

BREEDING The carrion crow is not colonial. Single pairs build an impressive and complex stick-nest in a treetop or pylon, sometimes on a cliff or building. The single clutch of three to six eggs is laid in early to mid-spring and is incubated for 18–19 days. The young leave the nest at the age of 28–38 days and become independent about a month later.

MIGRATION This is a sedentary bird.

ABUNDANCE It is very common.

◁ *Carrion crows often hide food away for retrieval at a later time. They often recover such food and then dunk it into water to wash it.*

CARRION CROW

△ *A fine of portrait of one of the garden's less popular birds. Note the hook at the tip of the upper mandible, which helps to cut up animal flesh.*

Crushing experience

Carrion crows in Sendai city, Japan, have recently become world-famous for a quite astonishing piece of ingenuity. Faced with a nearby supply of walnuts – which make a tasty meal, but are too hard for the birds to open with their bills – they have taken to using a form of brute force in order to open them. The brute force is not their own. The birds take the nuts to a traffic junction, where they place the nuts in the paths of waiting cars when the traffic lights are red and then retreat to a safe distance. The cars duly run over the walnuts and, once the lights have turned red again, the crows can retrieve their meal.

All change

The social structure of crows defines two strata of society: paired birds with a territory and non-breeding flocks. There is great pressure on the former from the latter because all birds aspire to a territory of their own and there are rarely enough territories to go round. Much aggression and bad feeling results from this.

The tension comes to a head when a member of a territory-holding pair dies. Without standing on ceremony, the remaining bird may be disinherited within hours and find its status changed. On some occasions the change may be even more jolting: the surviving bird remains where it is, but finds that almost immediately it has a new partner.

Hooded crow

Species: *Corvus cornix*
Family: Corvidae

◁ *The hooded crow is easily distinguished from its close relative the carrion crow by its two-toned plumage.*

▷ *Two hooded crows bicker over a piece of carrion. These birds will eat almost anything, animal or vegetable.*

IDENTIFICATION In shape and size, the hooded crow (45–47 cm/17¾–18½ in) is virtually identical to the carrion crow (see pages 136–137) but it has distinctive plumage, with a black head, wings and tail and a black throat with an uneven, streaky edge as it borders the dirty-grey breast and belly. Nape and mantle are the same colour as the breast, giving a two-colour pattern. Some hybrids with the carrion crow have more black streaks.

The *male* and *female* look alike.

The *juvenile* (early to late summer) is browner-tinged, with some speckling on the back.

SHAPE AND CHARACTER This is a large, powerful bird with strong legs and a thick bill with a down-curved upper mandible with a cutting edge – features shared with the carrion crow. It is usually seen on the ground, walking shiftily and sometimes hopping with both feet. It is quite predatory, and not as sociable as the rook (see pages 134–135) – it breeds as a pair and only forms moderate-sized flocks. Its flight is like that of the carrion crow, with slow, almost ponderous wing-beats. It flies on a straight course.

VOICE It makes angry caws, often in series of three. Some calls are said to be hoarser than those of the carrion crow.

HABITAT This is a bird of open country with scattered trees or clumps of trees, usually nesting in a treetop. It takes well to farmland and suburban and urban landscapes.

FOOD It is omnivorous, eating a great deal of carrion and sometimes patrolling roads for kills. It also feeds at rubbish dumps. Otherwise it takes insects from grassland, grain from fields and the odd small mammal, plus some eggs and nestlings. It feeds on the ground, with occasional forays to the treetops to catch caterpillars and other insects.

IN THE GARDEN It visits the ground and bird tables for a range of scraps, especially meat.

BREEDING Single pairs build a complex stick-nest in a treetop, or sometimes in cliffs, buildings and other elevated sites. The single clutch of three to six eggs is laid in early to mid-spring and incubated for 18–19 days. The young leave the nest at 28–38 days and become independent about a month later.

MIGRATION In many places (including Britain) the hooded crow is sedentary, but some northern populations spill south or south-west in the autumn, returning in late winter to mid-spring.

ABUNDANCE This is a common bird.

HOODED CROW

Fishing crows

In far northern Europe hooded crows have learned some basic human communication. When a fisherman drop a fishing line through a hole in the ice, the tugging of a fish that has taken the bait releases a coloured flag to tell the fisherman from a distance that a fish has been caught. Apparently, local hooded crows have learned what the appearance of the flag means and fly down to the fishing line, pulling it up with their bills until the fish is revealed and can be eaten.

The joy of six

Every birdwatcher knows the angry calls of crows. The calls are often in threes, and recently it has been shown that members of the crow family can count up to six.

Three's a crowd

Some crow territories don't contain just a single pair of birds, but three. In such aggressive birds this is a curious arrangement, especially given how high the territorial stakes are in this species. However, it seems that it's all down to persistence. A young male is usually the interloper and just stays around in the territory, repelling attacks by the incumbent bird until the latter simply gives up trying.

Recently it has been shown that the 'third bird' isn't just a spectator of life as a pair, but a participant too. The extra birds have been seen helping to feed the pair's young.

Golden oriole

Species: *Oriolus oriolus*
Family: Oriolidae

IDENTIFICATION This brightly coloured forest bird (24 cm/9½ in, the size of a blackbird) of continental Europe is rare in Britain. The male is unmistakable, while the female is still bright and exotic-looking. It has a red eye and quite a stout red bill.

The *male* is one of Europe's most beautiful birds, its body plumage a stunning yellow, with jet-black wings and a black tail with yellow tips.

The *female* is yellow-green above and lemon-yellow below, with fine but conspicuous streaks on the underside. Some individuals are much brighter, resembling the males.

The *juvenile* (mid- to late summer) is a duller version of the female, with fewer streaks below and a browner bill.

First-year males resemble the females.

SHAPE AND CHARACTER This looks like a sturdy blackbird with a shorter tail, but flies in a thrush-like manner, with elegant, swooping wing-beats and shallow undulations. When landing in trees, it sweeps upward at the last moment, like a woodpecker. It is shy and likes to stay concealed in foliage and, despite its bright colours, can be hard to spot.

VOICE Its song is a glorious, exotic whistle with a flutey quality: 'weela-wee-o'. Both sexes have a surprising hoarse squawk, like a complaining cat.

HABITAT This is very much a bird of the trees, inhabiting open woodland, parks and gardens.

FOOD Insects are its primary breeding-season food, including large caterpillars of the type avoided by most other birds. It feeds in treetops most of the time. It also eats berries in late summer.

IN THE GARDEN The golden oriole is an incidental visitor to gardens. It comes to berries in summer and autumn.

BREEDING It makes an unusual nest, which is usually placed quite high in a tree, some 12–18 m (39–59 ft) above ground and within 6–7 m (19½–23 ft) of the top of the canopy. The male is responsible for most of the construction. Three to four eggs are incubated by both adults (mainly the female) for 14–15 days and the hatched young remain in the nest for the same time again.

MIGRATION This bird is a summer visitor, arriving in mid- to late spring and departing for its winter grounds somewhere in sub-Saharan Africa from mid-summer to early autumn.

ABUNDANCE Locally it is common.

◁ *Although not as spectacularly plumaged as the male, the female golden oriole is still brightly coloured. Note the red bill.*

GOLDEN ORIOLE CONFIDENTIAL

▷ *A male golden oriole brings food to the young in the nest. Despite its coloration, this bird is secretive and surprisingly difficult to spot.*

Nest egg

The nest of the golden oriole is unusual in several respects. First is its construction, being rather hammock-like and suspended from the fork of a branch. It is woven into the rim of the branches making up the fork, but if this isn't possible the builder may weave together nearby branches to make a support. Construction of the nest takes a week, or even two.

Second, the list of materials used for the nest makes lively reading. Paper and string are regulars, as are snake skins and plastic. One nest was found to comprise entirely surgical dressings, and another aluminium strips. Pride of place should go, however, to the oriole that lined its nest with a 1,000-franc banknote – the person who found this nest relieved it of that particular material. Who says that birdwatching cannot, on occasion, be profitable?

Nest protection

Orioles use various means to defend their nests. The most obvious one is to keep themselves, and their structures, as inconspicuous as possible. Indeed, despite their bright plumage, these birds merge very effectively into the canopy of green-leafed trees, and their nests are extremely difficult to find. In the presence of predators, the sitting female crouches, and this is usually enough.

However, sometimes a little vigour is called for. Orioles often nest close to aggressive species such as fieldfares and shrikes, which violently oppose intruders. But orioles are quite capable of aggression themselves, especially when several neighbouring pairs team up against the danger. On a few occasions, a few strikes of the bill have been enough to kill a predatory bird that has attempted to attack an oriole's brood.

Starling

Species: *Sturnus vulgaris*
Family: Sturnidae

△ *The female starling has a pinkish base to the lower mandible of the bill, a pale ring around the eye and more spots than the male.*

▷ *A male starling in summer. Note the bluish base to the bill. Also typical of starlings generally is the metallic colouration of the plumage.*

IDENTIFICATION The starling (22 cm/8¾ in, smaller than a blackbird) is a lively, highly sociable black bird with a spiky bill, common in most gardens. It has strong, pink legs; its gait is a jerky walk or run, not a hop. It has long wings, but rather a short tail. Its plumage is metallic green/ iridescent purple, covered by varying numbers of small white or brown spots.

The *breeding male* (late winter to mid-summer) has a yellow bill with a bluish-grey base and few spots on its plumage.

The *breeding female* (late winter to mid-summer) has a yellow bill with a pink-brown base and more white spots than the male. It has a thin pale ring around the eye all year.

The *non-breeding adult* (mid-summer to mid-winter) has a dark bill and heavy and intricately spotted plumage, with white spots below, buff spots above. The female has larger spots than the male.

The *juvenile* (early to late summer) has a dark bill and its plumage is grey-brown all over, with a pale throat and light streaking on the underparts.

The *immature* (mid-summer to early autumn) may look a mixture of juvenile and adult plumage, with a grey-brown head and a dark, spotted body.

SHAPE AND CHARACTER The starling's short tail and spiky bill are distinctive. It has a jaunty walk and a habit of probing its bill into the ground. In flight its wings are somewhat triangular in shape and it flies fast on a straight course, with rapid wing-beats followed by short wing closures and the odd glide. When flushed, flocks of starlings don't head for cover, but fly upwards and often circle round before landing.

The starling is extremely sociable, usually seen in gangs of ten or more during the day and gathering in large flocks in the evening or at good food sources.

VOICE It is noisy, with a wide range of calls. Often heard when the bird is taking off is a buzzing 'cherrr', and its alarm call is a sharp 'chett'; it also gives a conversational 'tsoo-ee'. The hoarse call of the young is a common mid-summer sound. Its song (mainly late winter to late spring) is a lively, rambling set of chatters, clicks and whistles, with a life of its own. Within singing bouts the starling often makes imitations of other sounds. When singing, the starling often flaps its wings at the same time, apparently code that it is on the lookout for a mate.

HABITAT This is primarily a bird of grassland, which needs trees or buildings in which to nest. Gardens of all kinds are an ideal habitat.

FOOD It is specially adapted to feed on grassland; the jaw muscles opening its bill are strong, enabling it to open the bill *in situ*, revealing buried items such as insect larvae. It also catches insects in flight. In autumn and winter it takes much plant food, including berries and seeds.

IN THE GARDEN A good lawn free of pesticides helps this species. It comes to bird tables for bread, scraps and bones, and will readily use nest-boxes with an entrance hole of 5 cm (2 in). Place several such boxes close together.

BREEDING It nests in a cavity, usually a hole in a tree, but also in buildings or rocks. The usual clutch comprises five to seven light-blue eggs, incubated by both sexes for 12–15 days. The young leave the nest 20–22 days after hatching, but may beg noisily from the parents for days afterwards. Sometimes there are two broods.

MIGRATION In some parts of Europe (such as Britain) the starling is resident all year. It is a summer visitor to some northern and eastern areas, and is welcomed like the swallow when it arrives. Large numbers travel to Britain and lowland Europe to spend the winter.

ABUNDANCE This bird is very common.

Dumping on the neighbours

Female starlings are not quite cuckoos, but do sometimes lay eggs in the nests of their neighbours – it's a great way of multiplying the number of offspring. Build your own nest and lay your own clutch, for sure, but in the meantime lay a few extra eggs and dump them in the nest of another member of your species.

Not many bird species perform egg-dumping, or 'intraspecific brood parasitism'. Perhaps there are several reasons why the starling does. For a start, the species is colonial, so there are always nests in close proximity. Second, since it nests in the darkness of holes, perhaps it is difficult for an incubating starling to monitor how many eggs it has.

◁ *There's no absolute guarantee that this parent is feeding its own chick. Starlings sometimes deposit an extra egg into their neighbour's clutch and hope that they will foster it.*

Roosts

Starlings are famous for their habit of roosting in large concentrations, often involving thousands or even millions of birds. They gather together at dusk at favoured sites, which include low scrub, reedbeds and buildings. The birds fly in silently, but as they settle down they begin to chatter and make a din.

The roosts are impressive and conspicuous, but why do starlings roost in such large numbers? They might do so for warmth, but they don't huddle together and there doesn't seem to be a significant micro-climate within the roost; or to avoid predation – being part of such a large group reduces the statistical chance of becoming a victim. Another possibility is that roosting sites are limited.

In recent times the theory has gained ground that communal roosts are nothing more than information centres. Starlings get a chance to monitor their peers and assess the condition of birds that have fed in different areas within reach of the roost.

More than just sitting

It is tempting to think that the act of sitting on a clutch of eggs – incubating – must be one of the breeding season's gentler tasks. Since it seems to involve simply sitting on eggs, and turning them every so often, incubation is easily regarded as rather passive – even restful. Measurements do not back this assumption up. Incubation is energetically costly. The metabolic rate of an incubating starling is 25–30 per cent higher than a non-incubating bird.

△ *The spiky bill of the starling is an adaptation to foraging on the ground by making holes.*

Smell in starlings

Birds are not generally given much credit for having a keen sense of smell. The olfactory part of their brain is small, compared to that concerned with vision or hearing. But in recent years researchers have been surprised by just what birds can do with this sense.

One revelation was that starlings use their sense of smell to select green plants to add to the nest. Once the structure is complete, the males bring in fresh material, much of which has a strikingly aromatic quality. In a German study the birds brought in plants of more than 60 different species. The most widely held theory is that this plant material contains insecticidal compounds that help reduce the number of parasites in the nest. Blue tits replenish such plants when they dry out, but starlings don't seem to replenish it, and it is possible that the males are bringing it as a form of courtship material – aromatic compounds to attract females to the nest.

Anting

Starlings are among a distinguished set of garden birds that have been observed 'anting'. This is a strange form of body care, in which a bird uses ants to anoint its plumage, presumably making use of the noxious fluids that the insects secrete to act as an irritant to the bird's body lice and other pests. Starlings ant by taking an insect in the bill and rubbing it against the feathers. Recent research has shown that anting is instinctive. Young birds don't have to be taught it, but start anting regularly from the age of 40 days.

Imitations

Starlings have a lively, rambling song that rattles along at quite a pace and is a cheery contribution to the urban soundscape. But the most famous aspect is its tendency to imitate a wide variety of extraneous sounds, including tawny owls, buzzards, chickens, pheasants and golden orioles – together with cats, crying babies, car alarms and telephones.

Why should a bird with a complex song imitate other sounds? Well, it simply elaborates the repertoire of an individual, just as we might insert a quotation or reference into a piece of writing. Among starlings, a large repertoire is attractive to the opposite sex, and adding mimicry is one way of embellishing the song.

Tree sparrow

Species: *Passer montanus*
Family: Passeridae

◁ *One of the curiosities of tree sparrows is that they lack a separate male and female plumage, so this bird could be of either gender.*

▷ *A stunning portrait of a tree sparrow. Note the chestnut crown, black cheek-spot and smart appearance – all good distinctions of a tree sparrow.*

IDENTIFICATION The tree sparrow (14 cm/5½ in, slightly smaller than a house sparrow) is a typical small brown bird, with a broad seed-eater's bill. It is very neat, with richly striped wings and back. It has a red-brown cap, white cheeks with a black spot in the middle and a white neck-collar. The underparts are a rather clean buff-grey. Strangely, the sexes are alike, whereas male and female house sparrows are entirely different.

The *male* and *female* are alike. This bill is black in summer and has a yellow base in winter.

The *juvenile* (early to late summer) is like the adult, but the crown is duller, the cheeks greyer and the cheek spot less obvious. The base of the bill is yellow.

SHAPE AND CHARACTER The tree sparrow is small and neat, with a rounded head. The tail is slightly shorter than that of a house sparrow. It is noticeably better turned out than the house sparrow, which often looks 'dirty'. Its flight is direct and fast, without the undulations typical of most small birds. It is noisy and chirpy, living in small colonies and sometimes gathering into larger flocks. It frequently associates with the house sparrow, and the two species will form large mixed flocks at times.

VOICE It makes chirps like a house sparrow, but these are a little softer and higher-pitched. In flight it makes a sharp 'tett-tett' call, quite unlike the house sparrow.

HABITAT This bird is less inclined to occur in urban habitats than the house sparrow, but is often common around suburban gardens, parks and farmland.

FOOD It mainly eats seeds, taken from crops and low plants. It usually feeds on the ground, but will 'ride' cereal stems and take the seeds *in situ*. It feeds on insects in the breeding season and gives them to the young.

IN THE GARDEN The tree sparrow is shy, but where it is common will visit bird tables and hanging feeders for seeds and the odd scraps. It comes readily to enclosed nest-boxes, with an entrance hole of 2.8 cm (1 in).

BREEDING The tree sparrow breeds in small, loose colonies, nesting in holes in trees, buildings or rocks, sometimes among creepers or in a haystack. It lays five to six eggs and incubates them for 11–14 days; the young leave the nest after 15–20 days. It often brings up two, and sometimes three, broods.

MIGRATION This bird is mainly sedentary, but small numbers of individuals make short-distance movements.

ABUNDANCE The tree sparrow is common in gardens where it occurs, but has much declined in some places, including the British Isles.

TREE SPARROW

Multiple matings

The rate of copulation among bird species varies a great deal. Some copulate repeatedly, others just once. For male birds, the advantage of multiple coupling is obvious: by flooding a female with his sperm, a male can put the seal upon his paternity of a certain brood. In the dunnock, a bird in which males are sometimes forced to share a mate with one or more other males, it is not unusual for bigamous females to copulate over 500 times in the course of a season. In an atmosphere of competition with other males, that is the smart thing to do.

For females, multiple matings would appear to be anything but desirable. They are not necessary, take up time that could be spent feeding and expose both members of the distracted pair to potential predation. Yet why are the copulation rates of some birds so high?

Recent work on the tree sparrow has shown that repeated copulation enhances a female's overall reproductive performance. Scientists found that females that permitted higher copulation rates produced larger clutches and brought more fledglings into the world. Furthermore, their incubation periods were shorter than those that copulated less often.

A time for pairing

Tree sparrows (and house sparrows) are unusual among small birds in that they generally pair up in the autumn rather than the spring. This is probably related to their colony structure: new recruits to colonies – in the form of young birds – are admitted in the autumn, after the breeding season and moult, so it is a time for these newcomers to make their mark.

In early to mid-autumn all male members return to roost at the colony site and spend a short period of the morning singing near the nests, either at their old haunts or, in the case of recruits, at territories they hope to take over. Their singing attracts the females, and these mornings are the time when new pairs form. For the rest of the day the birds go off to feed elsewhere. By late autumn all morning singing sessions cease, but by then new partnerships are usually established and will remain intact until breeding starts the following year.

▽ *Tree sparrows are usually paired up by the winter, by which time they have joined up with an established colony.*

▷ *Recent research has shown that female tree sparrows benefit from multiple copulations – it enhances their overall breeding success.*

House sparrow

Species: *Passer domesticus*
Family: Passeridae

IDENTIFICATION The house sparrow (14–15 cm/ 5½–6 in) is the archetypal small brown bird, although it is slightly larger than most. Famously abundant, it is a scruffy, tame, but wary bird of built-up areas. It has a heavy seed-eater's bill. It is brown above, with richly patterned wings and back, and grey-brown below, slightly streaked.

The *adult male breeding* (late winter to mid-summer) has a grey crown and cheeks, separated by a red-brown stripe running from behind the eye to the nape. It has a small black patch from eye to bill, and a large black 'badge' on the breast, which varies in size. The bill is black.

The *adult male breeding* (late summer to late winter) has a much smaller breast badge, just a bib. Its head pattern is less sharp and its bill yellowish.

The *adult female* (all year) differs from the male in its head and breast pattern. It lacks the badge, or any black marks at all. It has a dull brown crown, grey cheeks and yellowish bill. Its most obvious feature is a pale line running from behind the eye towards the nape.

The *juvenile* (early summer to early autumn) is similar to the female, but looks washed out. It often shows the remains of a yellow gape at the base of the bill.

△ *The female house sparrow is the archetypal undistinguished small brown bird. Note the pale stripe behind the eye.*

▷ *House sparrows live in small, permanent colonies with their own communal territories. This is a male.*

SHAPE AND CHARACTER The house sparrow is quite robust for a small bird, with a large head and heavy conical bill. It flies with unusual directness and continuous wing-beats, without the undulations of other small birds. It is a perky, noisy and tame species that is nonetheless wary of people. It has a tendency to look scruffy, drooping its wings and ruffling its feathers. Males hold up their tails and spread their wings in display. The sparrow is extremely sociable, living in moderate-sized colonies.

VOICE It has a variety of unremarkable cheeps and chirps.

HABITAT This bird is very much found in the company of people, most frequently in cities, towns, suburbs and farm buildings.

FOOD It is primarily a vegetarian, feeding on seeds from low plants and cereals. It also takes berries, buds and shoots, including those of crocuses and forsythias. It usually feeds on the ground, but can perch on plant stems, and on occasion will even catch flying ants and other insects in the air. Insects are its principal food during the breeding season and are given to chicks in the nest.

IN THE GARDEN Artificial feeders are its natural habitat! It comes abundantly for all kinds of scraps, especially seeds and cereals. It will take to enclosed nest-boxes with an entrance hole of 3.2 cm (1¼ in). It has a habit of monopolizing boxes.

BREEDING The house sparrow usually sites its nest in a hole on the outside of a building, but occasionally builds a free-standing structure in a tree. The clutch consists of three to five whitish eggs with variable blotches, incubated for 11–14 days. The parents bring in insects to the young, which fledge at 11–16 days. They are fed outside the nest for another two weeks.

MIGRATION This bird is sedentary, but may make local movements in late summer.

ABUNDANCE A very common garden bird, it has declined recently in some places.

HOUSE SPARROW

Badges of honour

One of the most conspicuous features of male house sparrows, as opposed to females, is the black bib on the throat. The size of the bib differs between individuals, and experiments have shown that it is an expression of dominance. Males with bibs of 4–5 sq cm (½–¾ sq in) are dominant over those with smaller bibs wherever there is an aggressive encounter, such as at a feeding tray.

But is it possible for a bird to grow a greater bib size and thus cheat its way to dominance? At first this seemed likely, because scientists in Denmark found no relationship between the size of the bib and other physical characteristics, such as body size, age or nutritional condition. However, when they introduced birds to a stable flock that had had their bibs enlarged experimentally by dyeing, they found no change in the outcome of conflicts. The genuinely large-bibbed males kept on winning all their encounters, putting the interlopers in their place.

This shows that bib size in house sparrows is a genuine reflection of pure fighting ability. Birds, if you like, earn their badges, and if anyone tries to cheat he is in for trouble.

House sparrow 'summer break'

House sparrows are among the most sedentary of birds, living for much of their lives in one place, along with other members of the same colony. Usually the only time they move is after leaving the nest and finding a colony to join for the first time.

However, every year the house sparrow effectively has an annual 'holiday'. It happens in late summer when the grain in arable fields is ripening and there is a glut of food. For a few short weeks the birds abandon their colonies and go on a local binge, never moving more than about 2 km (1¼ miles) away, but nevertheless staying away from home.

◁ *There is a direct correlation between the size of the bib and a male's fighting ability.*

▷ *In the late summer, large numbers of house sparrows gather at agricultural fields where crops are ripening.*

A period of refraction

In common with many garden birds, house sparrows have a specific breeding season. This season is under hormonal control and therefore the birds cannot 'escape' it and breed opportunistically. The end of the breeding season is dictated by the declining length of days in the temperate zone and induces important changes in the birds' physiology, including a regression in the size of the gonads; the bird then enters its so-called 'refractory period'.

Only when the days begin to lengthen do the sexual organs regrow. And, in the case of the house sparrow, they may increase in weight several hundredfold before breeding can take place again.

Infanticide

The house sparrow has recently joined the list of species in which infanticide has been recorded (see also page 43). It happens when one member of a pair is bereaved. If this occurs during the breeding season, the desperate individual may kill the young of a neighbouring brood. The rupture of family life may split up the pair and, incredibly, the killer may then pair up with the recently bereaved and salvage its own breeding season.

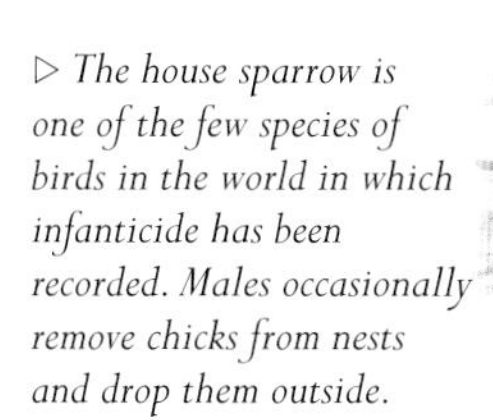

▷ *The house sparrow is one of the few species of birds in the world in which infanticide has been recorded. Males occasionally remove chicks from nests and drop them outside.*

Chaffinch

Species: *Fringilla coelebs*
Family: Fringillidae

IDENTIFICATION The chaffinch (14.5 cm/5¾ in) is the size of a sparrow, but has a much slimmer build with a peaked crown. This is a colourful songbird with a long tail with white sides and a complex wing-pattern: white shoulders, a Y-shaped central wing-bar and yellowish edges to the flight feathers. The rump is green. The plumage is not streaked at all. The bill is quite long and not as broad as many other seed-eaters' bills.

The *adult male breeding* (mid-winter to mid-summer) is very smart, with a blue-grey cap (black on the forehead), pink cheeks and breast, and a reddish-brown mantle. Its bill is grey.

The *adult male breeding* (late summer to early winter) is markedly less colourful, with a plum-pink breast and a subdued greyish-brown on the crown and nape. Its bill is paler.

The *adult female* is an anaemic light-brown version of the male. Its crown, nape and mantle are grey-brown, its cheeks grey and its underparts a dull greyish-white.

The *juvenile* (early to late summer) is similar to the female, but still paler.

SHAPE AND CHARACTER This is a slim songbird with a long tail. It often perches more horizontally than other finches, with its tail drooped down. It typically feeds on the ground, shuffling with short steps and nodding its head slightly, like a chicken; it also hops normally. Its flight is light and free, with more gentle undulations than other finches. It is able to swoop and hover to catch insects in flight. It is sociable only in the non-breeding season, when it may gather in large flocks.

VOICE The chaffinch has a bewildering vocabulary of calls, including a cheerful 'pink-pink'. Its flight call is 'choop', while a common perched call is 'huit!', often repeated endlessly. The song (late winter to early summer) is highly distinctive, a cheerful, accelerating rattle with a flourish at the end. The phrase is repeated again and again.

HABITAT This is a woodland bird that has adapted to parks and gardens, and is often very common.

FOOD In contrast to other finches (except the brambling), it feeds almost entirely on insects (chiefly caterpillars) in the breeding season and feeds these to its young. The rest of the year it feeds on a wide variety of seeds, including beech mast in the autumn, collecting it from the ground.

△ *Female chaffinches are duller and less contrasting than males, but note the distinctive Y-shaped wing-bar.*

▷ *The male chaffinch has a pink breast, blue-grey top of the head and smart brown back.*

IN THE GARDEN The chaffinch is a common visitor to bird tables and hanging feeders with perches, for nuts and seeds and other scraps. It often feeds on the ground, taking spilt grain from above.

BREEDING It begins breeding in late spring, building a neat, compact cup-nest. The usual clutch is four to five light-blue eggs, which are incubated for 10–16 days by the female only. The young leave the nest about 13 days after hatching and are fed out of the nest for about three weeks. There are often two broods.

MIGRATION Many chaffinches are non-migratory, but the northern and eastern populations evacuate their breeding grounds from early to late autumn to winter further south, in the British Isles southwards to the Mediterranean. They return between late winter and late spring.

ABUNDANCE The chaffinch is a very common bird.

CHAFFINCH

Single-sex flocks

The chaffinch demonstrates a trait known as 'differential migration' – a grand term to say that males and females (as well as juveniles) exhibit different migrations. In many species, males do not migrate as far south as females and juveniles in the wintertime, partly because they are better able to withstand the cold, and partly because they wish to remain in pole position for a rapid return to the breeding grounds – and thereby lay claim to a territory as quickly as possible. Since the females don't need to set up territories, they aren't in such a hurry to return.

In the chaffinch it is typical, in winter, to find flocks consisting wholly or mainly of just one sex. Most famously, the man who gave the chaffinch its scientific name of *Fringilla coelebs* was Carl Linnaeus. Where he lived, female chaffinches evacuated the area in winter, leaving the males behind – and the term *coelebs* means 'bachelor'.

Territory size

It is difficult to measure the size of any species' territory, since individual spaces vary enormously – for example, among chaffinches, territories in food-poor areas are six to eight times larger than they are in food-rich areas. Nevertheless, as a guide, chaffinches in Finland have an average territory size of 4,000 sq m (43,000 sq ft).

▽ *Chaffinches are extremely sociable birds outside the breeding season. In winter, flocks are often dominated by one gender.*

◁ *In spring the male chaffinch delivers a cheerful song that is an accelerating phrase with a flourish at the finish. Males in different places deliver a slightly different suite of songs – they have dialects.*

Song patterns

The chaffinch has a simple song that has been a useful tool in the study of bird vocalizations, and it exhibits a number of tendencies that hold for many other species.

One is that male chaffinches learn their songs. If reared in isolation, they are unable to sing normally – they need to hear a chaffinch song in order to match it. In youngsters there are two peaks of learning: when they fledge, but before the adults stop singing, in summer, and when the adults start being territorial again the following spring. As a result of this learning method, it is perhaps not surprising that chaffinches have dialects. Birds in the wild learn from the singers around them and are brought up to copy local tendencies and variations.

When territorial singing commences in the spring, there is much conflict between males, but this quickly resolves into what might be described as an uneasy truce. Birds learn to recognize the songs of all their neighbours and, on hearing the familiar, they no longer exhibit aggressive behaviour. However, the moment an intruder appears there is an instant violent reaction, proving that the birds know that a new bird is in town.

Catch a falling seed

The chaffinch has the most wide-ranging and most varied diet of any member of the finch family, but it is almost as interesting for what it cannot do when foraging as for what it can.

Unlike most finches, the chaffinch takes a large number of insects in the spring and summer and, using its long wings and tail, is capable of making short fly-catching flies into the air, and even hovering in front of caterpillars hanging from silken threads before snatching them. It has been recorded catching falling pine seeds in flight and, conversely, has also been seen to drop seeds onto a hard surface to try to open them. However, in contrast to other finches, the chaffinch simply cannot tolerate feeding upside-down (goldfinches and siskins, for example, do this all the time) and doesn't seem to be able to use its feet when handling food (unlike a crossbill, for example).

All-purpose territories

Chaffinches have a profoundly different breeding system to all other finches except bramblings and it's due to the fact that chaffinches feed their young on insects rather than seeds.

Insects are so abundant in the trees that a chaffinch can easily defend a territory that provides for all a pair's needs, separate from any other birds. This is impossible for the finches that provide seeds for their young; seeds are widely dispersed and it would be impossible to defend an area providing enough of them. So most finches are sociable in the breeding season, often foraging together and nesting in loose aggregations. Chaffinches are territorial and have a distinctive song rather than a rambling one.

Brambling

Species: *Fringilla montifringilla*
Family: Fringillidae

IDENTIFICATION The brambling (14 cm/5½ in) is fractionally shorter than the chaffinch (see pages 154–157) with a shorter tail, although it is more strongly forked. This colourful woodland finch has a distinct orange wash to its plumage at all times, and a slightly heavier bill than the chaffinch. It has a similar wing pattern to the chaffinch, but its shoulders and wing-bars are orange. It has a white rump.

The *adult male breeding* (mid-winter to mid-summer) is very handsome and unmistakable. It has a completely black head and back, and an orange (apricot) throat and breast.

The *adult male breeding* (late summer to mid-winter) has colourful plumage clouded by pale tips to the feathers, especially on the head.

The *adult female breeding* (late winter to mid-summer) is quite dull, with a mottled brown crown and back. Little orange shows on the wing or breast.

The *adult female non-breeding* (late summer to mid-winter) is similar to the non-breeding male, but the head is much greyer and the orange is not as bright, especially on the shoulders.

SHAPE AND CHARACTER In many respects the brambling is similar to the chaffinch, but has a slightly stouter bill and a shorter, more forked tail. These, together with the larger head, show up in flight, which is light and bouncy. It is very sociable and is often seen in large, rather nervous flocks in the winter.

VOICE Its flight call, often heard on migration, is a gentle 'yeck'; when perched, it may give a buzzing 'zhwee!' like a small parrot. Its song (late winter to mid-summer) is very simple, a repeated wheeze – flatter and more even than the greenfinch's song.

HABITAT It is found in open woodland in the breeding areas and typically at the woodland edge in early winter, especially in beech woods. Later on it will visit farmland, parks and gardens for other seeds.

FOOD In common with the chaffinch, the brambling feeds almost entirely on insects (caterpillars, beetles) in the breeding season. For the rest of the year it relies on seeds (especially beech), later on switching to other foods. It forages on the ground and can dig through snow with its bill.

IN THE GARDEN It comes sparingly to bird tables and ground feeders for seeds.

BREEDING As a breeding bird, the brambling is restricted to Scandinavian and Russian forests. It builds a cup-nest, often quite high up in a bush or tree. The clutch of five to seven eggs is incubated for 11–12 days and, once hatched, the young remain in the nest for 13–14 days. There are usually two broods.

MIGRATION The brambling is highly migratory, evacuating its breeding areas in early to mid-autumn and returning in late winter to late spring. In most of central Europe (including Britain) it is a winter visitor.

ABUNDANCE It is generally uncommon in gardens.

◁ *The brambling is similar to its close relative the chaffinch, but has a distinctly orange wash to the plumage. Note the yellow bill, typical of birds in winter.*

▷ *A male brambling almost in breeding plumage, with its head beginning to turn jet black.*

BRAMBLING

◁ *Bramblings tend only to be occasional visitors to gardens. If they do come, it is usually in the company of chaffinches.*

Short cut to good looks

All birds need to change their feathers at least once a year in order to survive, a process that takes place during the autumn: the annual moult. Many birds also have a second moult in the early spring, in order for the breeding plumage to look at its best and advertise its wearer. Moult is an exhausting process – it could be compared to the growth spurt of adolescence in humans, when the individual seems devoid of extra bounce. So in an ideal world a bird would acquire a new look in the spring without actually moulting; that would save a lot of effort.

It happens that bramblings (and other finches) have hit upon a way of doing exactly that. This is known as 'abrasion'. Essentially the birds undertake their moult in the autumn as usual, growing new feathers that are dull at the tips, but bright just below. On the brambling's head, for instance, the fresh feathers have brown tips, but further down they are black. Over time, the feathers wear roughly equally over the plumage. The dull tips are worn off (abraded) and reveal the brighter colours below. The effect is an illusion: the colourful plumage looks new and smart, but is actually the result of wear – it works nonetheless.

Reverse migration

Sometimes human observers become so awestruck by the phenomenon of migration that they forget that birds sometimes make mistakes. Occasionally an individual flies in the right direction, but goes too far, arriving in conditions that are not suitable for it. In such a situation, it might pay the traveller to turn around and fly straight back on the bearing from which it came. When this occurs it is known as 'reverse migration'. This phenomenon has been studied in bramblings on their mass migration down the coast of Sweden. Researchers found that birds frequently undertook these movements, often a few hours after their original migratory departure in the morning. As might be expected, those that reversed were generally in poorer condition than birds that ploughed on.

Floating bramblings

It appears that, if you are a brambling, all is not lost on your migration, even if you have to pitch onto the sea. Individual finches have been observed floating on the water surface, only to fly up again, apparently unharmed, when disturbed by a boat.

△ *Migrating bramblings occasionally ditch into the sea but, amazingly, this isn't always fatal, and they have been recorded flying off from the surface unharmed.*

◁ *Bramblings are extremely sociable and their winter roosts may attract thousands – or even millions – of birds.*

Greenfinch

Species: *Carduelis chloris*
Family: Fringillidae

IDENTIFICATION The greenfinch (15 cm/6 in, about the size of a house sparrow) is a heavy-bodied, large-billed finch that is common in gardens. It is distinguished by its thick, ivory-coloured bill, yellow wing-bar (along the edge of the closed wing), yellow tail-sides and generally green plumage.

The *adult male breeding* (mid-winter to mid-summer) is a pleasing apple-green colour, with a bright, large yellow wing-panel and tail-sides.

The *adult male breeding* (late summer to late autumn) is greyer than the summer plumage; grey-brown above, with greyer cheeks.

The *adult female* (all year) is not as bright green as the male, with a browner, slightly dark-streaked mantle and rather little yellow on the tail.

The *juvenile* (late spring to late summer) differs from the adults by being heavily streaked on the underparts and mantle. It has small yellow patches on the tail and wing.

△The thick bill allows the greenfinch to exploit many food sources that are unavailable to finches with thinner, weaker bills – such as rose hips, for instance.

SHAPE AND CHARACTER This is a large finch with a very heavy head and bill, plus a decidedly short tail. It flies powerfully with big undulations and swoops. When perched, often high up in a tree, it frequently takes a vertical posture. It feeds on the ground and on hanging feeders. It is sociable at all times of the year. The greenfinch has a spirited display flight in the spring: it flies with elegant, full wing-beats at treetop height and describes a figure-of-eight or a circle, pitching from side to side.

VOICE Its constantly uttered call is a 'chup-chup…' with many variations, delivered both in flight and perched. It also makes a questioning, buzzing 'do-eee'. Its song is a series of trills at different paces, often interspersed with an emphasized, drawn-out 'dzhweee'.

HABITAT The greenfinch lives in a variety of bushy and scrubby habitats. Gardens are much to its liking.

FOOD It is very much a seed-eater, taking small numbers of insects in summer. It eats a wide variety of seeds throughout the year, including groundsel, chickweed, sunflower and yew, the last being separated from the berries, whose seedcoat is poisonous. It usually feeds on the ground, but also on herbs (such as sunflowers) and on trees.

IN THE GARDEN This bird is a relentless monopolizer of hanging seed and nut-feeders, especially those containing black sunflower seeds and peanuts. It will also feed on sunflower heads.

BREEDING Pairs often form in winter flocks. Four to six eggs are incubated for 12–14 days and, once hatched, the young leave the nest 13–16 days later. They cannot always fly well upon fledging, and may be fed and looked after by the male while the female starts the next clutch. Some pairs have three broods.

MIGRATION The greenfinch is resident over most of Europe, but this masks the fact that many individuals make south-westerly movements in autumn and return in spring.

ABUNDANCE It is a very common bird.

Providing for the future

Recent studies have revealed that the greenfinch does not always pair up in the conventional way. Rather than a single male joining forces with a single female, it is common for male birds to have more than one mate: two or even three. About 25 per cent of males are polygamous in this way.

Whenever this happens in an even sex ratio, it follows that some male birds will pass the breeding season without acquiring a mate at all – there simply aren't enough females to go round when some males monopolize them. Such males don't just resign themselves to a summer of idleness, but actually pitch for their future. They do this by bringing food to an attached female's offspring. It won't help them in the short term, but it might persuade a female to accept them as a mate during the following breeding season.

A multi-purpose bill

The bill of the greenfinch is thicker than that of most other finches. This means that it has the tool to eat from a wider range of plants, especially the harder sort. It is often seen in places where other finches do not or cannot go: on the seedheads of sunflowers, yew berries and blackberries.

The sheer adaptability of its bill is reflected in the statistics. In one study the greenfinch was seen to take seeds from every one of a location's most common plants. And over a wide area it has been recorded feeding from hundreds of different plant species.

▽ *The pink bill, yellow wing-bar and slight 'frowning' expression all help to distinguish the greenfinch from similar species.*

Goldfinch

Species: *Carduelis carduelis*
Family: Fringillidae

IDENTIFICATION This glorious, brightly coloured songbird (12 cm/4¾ in, smaller than the chaffinch) is common in gardens. It has a sharp, conical, ivory-coloured bill. Its unique tricoloured head is crimson on the face, with a white and then a black band behind. Its main plumage colour is sandy-brown, but the chest is white. The wing is black with a brilliant yellow wing-bar across it, and white spots on the tips of the flight feathers; its tail is black with white spots.

The *male* has a little more red on the face than the *female*.

The *juvenile* (early summer to early autumn) has wings that are much the same as the adults', but its head lacks any bold coloration – and it looks a little odd, just brown and faintly streaked.

SHAPE AND CHARACTER The goldfinch is a small, delicate songbird that is usually seen in flocks feeding on herbs, especially thistles. It has a light, fast, undulating flight. Flocks pack closely in flight, seemingly connected by threads that rein them in. When feeding, it is highly agile, capable of hanging almost upside-down, flapping its wings to keep its balance. It is sociable throughout the year.

VOICE Its main call is a cheerful 'tick-lit' and variations. Its song (early spring to mid-summer) is a liquid, trickling ramble of those same calls, interspersed with harsh notes.

HABITAT It lives anywhere tall weeds occur, in farmland, gardens, parks, churchyards and the edges of woods.

FOOD The goldfinch feeds almost entirely on seeds, most of them from thistles (which can comprise 60 per cent of its annual diet), but also on dandelions, burdocks and other related plants, teasels and lavender. It takes a small numbers of invertebrates in the summer.

IN THE GARDEN It has recently become a fixture in many (especially British) gardens, attracted by nyjer (niger) seeds in hanging feeders. You can also try pouring these seeds into dead teasel heads.

BREEDING The goldfinch often breeds late, in clusters of pairs known as 'neighbourhood groups' – up to nine pairs in all. It has an unusual nest site, typically in a tree right at the end of a branch, often over water, and therefore inaccessible to ground predators. The female lays four to six eggs, which are incubated for 11–14 days. Once hatched, the chicks are fed by both adults and leave the nest at 13–18 days. There are two or even three broods a year.

△ *It's unusual for a garden to be visited by a single goldfinch – instead, they usually come in small groups.*

▷ *The brilliant coloration of the goldfinch makes this bird unmistakable. Male and female look much alike.*

MIGRATION It is resident, but many individual birds move south and south-west in the autumn, returning to the breeding areas from late winter onwards.

ABUNDANCE Locally the goldfinch is common.

Males and teasels

Many birdwatchers delight in seeing goldfinches visit their gardens for teasel seeds – it is the only species that can deal with the tough seedheads. What they often don't realize is that they are watching a good example of food partitioning among the sexes. The birds successfully opening the bracts and extracting the seeds are probably all males.

The female's bill is, on average, 1 mm shorter than that of the male. It might not sound much, but it means that females must work much harder on teasels to get anything – turning down the surrounding spikes. They can only obtain food at a rate of one seed to every four that a male can cope with, and frequently they don't bother. If the males work the teasels, that leaves more thistles available for the females.

◁ *It is usually only male goldfinches that tackle the seed-heads of teasels. The females have shorter bills.*

▷ *Our jaws are designed to bite, but the stronger muscles in the goldfinch's jaw open the bill. This helps the birds to pry open seedheads.*

△ *Unusually for small birds, goldfinch nestlings are fed mainly on regurgitated seeds, with only a few insects.*

Bills for a purpose

The goldfinch is a thistle specialist and its bill has several adaptations for this purpose. First, the bill is, for a finch, quite thin and narrow, ideal for inserting into the tight spaces between the bracts of plants.

Second is the bill's musculature. Human jaws are adapted for biting: all the muscular strength is devoted to closing the mandibles together. In the goldfinch and some of its relatives it is the opposite: the muscular strength goes into opening the bill so that, when inserted into a tight space, it makes the space wider so that the tongue can scoop the seeds out.

Nest fit for purpose

The nest of a goldfinch is expertly made. It is tightly constructed – in the case of one of its closest relatives, the American goldfinch, it can actually hold rainwater when empty. Building a nest at the far end of branches means that the site is usually safe from predators, but somewhat at the mercy of the wind. As a result, goldfinches build up a particularly high rim, so that if the wind is very strong, the contents don't fall out in a sudden gust.

The working day

A study of goldfinches has given a fascinating insight into the number of eggs laid by similar, or the same species, in different parts of the world.

Goldfinches were introduced by people into Australia just over 100 years ago. Since they have settled into the new continent, their clutch size has got smaller: they lay an average of 3.7 eggs per clutch, whereas birds in Britain, its ancestral population, lay an average of 5 eggs per clutch – significantly more.

Why? The main reason seems to be the different day lengths in the two continents. It seems that the 'working day' of the feeding adults affects the number of young that they can raise in a brood.

Siskin

Species: *Carduelis spinus*
Family: Fringillidae

◁ *Siskins are distinguished by their two yellow wing-bars and heavily streaked flanks. This is a female.*

▷ *The male siskin has a coal black crown and chin.*

IDENTIFICATION The siskin (12 cm/4¾ in, slightly larger than a blue tit) is an acrobatic tree-dwelling finch, smaller than most other garden seed-eaters. It has a very thin but pointed bill. Overall it is yellow-green, with two yellow wing-bars (one at the shoulders) across the closed wing; the tail-sides and rump are also yellow. It is heavily streaked on the flanks.

The *male* has a coal-black crown and chin, a yellow rump and yellow on the face.

The *female* lacks the dark crown and chin, and is greener on the face. Its rump is streaked.

The *juvenile* (early to late summer) is much paler – whitish – and its wing-bars are not as bright. It is heavily streaked on the back.

SHAPE AND CHARACTER This is a small-headed finch with a short tail and a sharp, small bill. Its flight is fast and light, with flock members keeping tightly together. When disturbed, flocks sometimes 'explode' out from treetops all at once. The siskin is very acrobatic and has a distinct tendency to feed upside-down, presumably a habit acquired by working conifer cones. It is extremely sociable.

VOICE It has several calls, including the flight call – a soft, somewhat feeble and creaking 'tseeu'. It also makes a sparrow-like chatter. Its song is a sizzling medley of fast notes and twitters, interrupted now and then by long buzzing trills that sound like flatulent bursts. In common with other finches, it often sings in chorus.

HABITAT This is a woodland bird that breeds in conifers (mainly spruce) and usually feeds on alder seeds in the winter. In many ways it is an unusual garden bird, but in some areas has made a habit of visiting hanging feeders when the alder-seed crop is exhausted.

FOOD The siskin eats almost entirely seeds, plus a few invertebrates in the summer. It primarily takes spruce, pine, alder and birch seeds, with a few from herbs such as docks.

IN THE GARDEN It comes mainly to hanging feeders for black sunflower seeds and nyjer, often in small parties. It also values drinking water nearby.

BREEDING The siskin doesn't normally breed in gardens. It builds its nest in a conifer, usually at the end of a branch well above ground. The clutch of three to five eggs is incubated by the female for 12–13 days. The young, fed on seed paste by regurgitation, leave the nest after 13–17 days.

MIGRATION Most northern European birds are migratory, leaving their breeding areas in late summer to mid-autumn and wintering south within Europe (including Britain). Some siskins are resident. The migrants return in late winter to mid-spring.

ABUNDANCE This bird is local.

SISKIN CONFIDENTIAL

Food offerings

It is common among birds for males to pass food to females in the breeding season. Known as 'courtship-feeding', this is a male's way of helping its mate prepare for the rigours of breeding. By offering the female food it has not found by itself, the female is able to save its energies.

However, a peculiar ritual has been observed among siskins in spring and summer: males offering food to males! At first this seems inexplicable, but it is allied to the birds' sociable nature. Siskins usually breed in small groups, with several males present. Early in the breeding season, skirmishes often break out between rivals and, as it becomes clear who is dominant, the losing bird appeases the conqueror by passing him food, bill-to-bill. As the season progresses the victor requests renewed offerings, and in this way further fights are avoided.

Reading the plumage

In common with many birds, the siskin displays on its plumage various 'badges': features that indicate its qualities as an individual. Two in particular have caught the attention of scientists: the black bib of the male and the bright-yellow central wing-bar.

The size of the bib is directly related to a bird's social dominance within the flock. Birds with large bibs (produced by the pigment melanin) are dominant over those with small bibs – a difference that is easy for the birds to see in social encounters.

What of the yellow wing-bar, though? Scientists researched whether the length of the bar (created by carotenoid pigments from the bird's diet) might have a relationship to a bird's foraging ability. They tested this using two measures: how often a bird calls when isolated, and how often it responds to the presence of live decoy siskins near a feeding area. The rate of calling and the rate of response to decoys would measure how much a bird depends on others to find food, and hence its ability (or lack of it) to be adaptable when foraging. Fascinatingly, they found a link. Birds with long yellow wing-bars responded less to decoys than those with shorter wing-bars, and called out less when isolated. Such birds would be more self-reliant when foraging.

▷ *In flocks there is always a strict hierarchy of individuals. It is maintained by subordinate birds bringing food offerings to dominant birds.*

Serin

Species: *Serinus serinus*
Family: Fringillidae

IDENTIFICATION The serin (11.5 cm/4½ in, the size of a blue tit) is a small, effervescent continental European finch with an extraordinarily stubby bill that seems too short for it. It is heavily streaked above and below, with narrow (barely noticeable) yellow wing-bars. Its rump, however, is bold yellow and conspicuous. Its tail is mainly dark. It perches high, but feeds mainly on the ground.

The *male* is a bright lemon-yellow on the breast, throat and head, and white on the flanks.

The *female* is a paler yellow-green on the breast and head. Its rump is not so bright.

The *juvenile* (late spring to late summer) is a brown version of the adult, without any yellow.

SHAPE AND CHARACTER This bird is distinctive, with its minute bill and large head. It flies freely, with steep undulations that are also a good clue. It is very restless. It feeds frequently on the ground, in small flocks, and less often in trees than some other finches. It has a splendid song-flight, launching into the air and progressing with slow, deep wing-beats in erratic loops and whirls, pitching from side to side, then circling back down.

VOICE Its main (flight) call is a silvery 'trilli-lit', which is quite distinctive. Its fizzing song is like the jingle of a very small set of keys, shaken very quickly.

HABITAT This is a forest-edge bird that has fitted well into the suburban scene where there are clumps of bushes and trees. It likes sunny places, and is often found in orchards, churchyards and parks. It makes good use of overhead wires, often singing from them, and frequently nests in conifers in gardens.

FOOD The siskin eats small seeds, including those of grasses. It forages on the ground and also on top of herb seedheads. It sometimes eats in trees, mainly in early spring.

IN THE GARDEN It will come sparingly for seeds, usually at ground stations.

BREEDING This bird selects a nest site in a bush or small tree; conifers are preferred. The clutch is of three to four eggs, and many pairs bring up two broods. The eggs are incubated for 13 days. Once they hatch, the youngsters remain in the nest for 14–15 days and leave the adults' care after another week or so. The siskin does not breed in the British Isles.

MIGRATION Many populations are sedentary, especially in southern Europe. Northern birds usually head south-west in early to late autumn, not going far (to North Africa at most). They return in late winter to late spring.

ABUNDANCE This is a common bird.

◁ *Serins are small streaky finches with a remarkably short, stubby bill. This is a female.*

SERIN CONFIDENTIAL

△ *The effervescent male serin has a delightful fizzing, rushed song, often delivered from a high perch or overhead wire.*

Song stimulation

The song of a male bird has a number of functions: it delineates territory and keeps males away; it also acts as self-advertising to attract females to a breeding territory. A third function, far less well known, is the physiological and behavioural effect that it has on females.

Experiments on serins have shown that listening to the mate singing has a direct effect on nest-building. Scientists took two sets of serins and, while they played recorded songs (in addition to the natural singing of the mate) every day to one set of females, the others were left alone. The birds bombarded with male song spent significantly more time nest-building than the ones that were not.

The peak of male serin song in the wild has been linked to the beginning of the growth of follicles in females, suggesting that the song also helps to regulate breeding in the mate.

Bright and healthy plumage

Scientists have discovered that the brightness of a male serin's plumage is directly related to the number of parasites it has on its plumage at the time of moult in the autumn – and this is an indicator of overall health. When male serins had their plumage experimentally treated with insecticide just before the moult, their plumage turned out brighter than that of the untreated control birds. A year later the same birds were left alone and their plumage turned no brighter than the controls, thus proving that it was the treatment that made the difference.

Bullfinch

Species: *Pyrrhula pyrrhula*
Family: Fringillidae

◁ *Although superficially similar to the chaffinch, the bullfinch has much bolder colours, with a black cap, grey back and cleaner wing-bar.*

▷ *Male and female bullfinches have a close and intimate pair bond. The male (below) has a much brighter breast colour than the female.*

IDENTIFICATION The bullfinch (15 cm/6 in) is the size of a chaffinch, but more heavily built. It is easily identified by its intense colours and simple, bold pattern. Both sexes have a neat black cap, black wings with a white wing-bar and a black tail. Look out for the white rump when the bird flies away.

The *male* has an intense strawberry-red breast and cheeks, and a grey back.

The *female* has a plum-coloured breast and cheeks, and a browner back than the male.

The *juvenile* (early to late summer) lacks the black cap of the adults.

SHAPE AND CHARACTER This is a burly finch with a thick bill that forms a curve with a crown; it is bull-necked. It exhibits different behaviour from other finches: it is quiet and not especially sociable, and is easily overlooked. It is often seen in pairs or groups of about six to ten. It has a quite powerful flight, rather similar to that of the chaffinch (see pages 154–157).

VOICE The bullfinch is not noisy or conversational like other finches. Its main call is a quiet whistle with a downward inflexion: 'pew'. Its rarely heard song (late winter to late spring) is like a sign creaking in the wind.

HABITAT This is really a bird of woodland, which uses gardens in winter and spring. Bullfinches are also found in scrub.

FOOD It is a seed-eater like the other finches, but also feeds a great deal on buds and shoots, with a few flowers and berries as well. It takes seeds of ash, dandelions and other weeds on the ground, plus large numbers of buds of fruit trees and forest trees. It eats more invertebrates than other finches (except the chaffinch and brambling) in the breeding season.

IN THE GARDEN These days the bullfinch is quite a frequent visitor to bird feeding stations, lured in by black sunflower seeds and hearts. It is also well known for eating the buds of forsythia, fruit trees, birch, and so on.

BREEDING It builds a nest in thick bushy vegetation, often surprisingly low down. The female lays four to five eggs, which are incubated for 12–15 days. The young are fed by both parents and leave the nest 14–16 days later, but are still looked after for as long as 20 more days. Despite this, some pairs bring up two broods, and even three.

MIGRATION This bird is basically resident. Some individuals move south in mid- to late autumn and return in late winter to mid-spring, but in much of the range bullfinches can be seen all year.

ABUNDANCE It is a fairly common bird.

△ *Once uncommon at feeders, bullfinches have become more regular recently, perhaps because the environment outside the garden provides less food for them.*

Pouching food

In the breeding season, bullfinches form special pouches in the floor of their mouths – organs that are not present in winter. Each pouch opens on either side of the tongue and they are used as storage organs to enable the birds to take large meals back to the young waiting in the nest. When the bullfinch's pouches are full, with 1 cubic cm (0.06 cubic in) or so of material, they extend back under the jaw as far as the neck, so that a bird carrying a full load shows a bulging throat.

The pouches are clearly designed to reduce the number of visits the adults need to make to the nest. They only visit about once every half-hour or so, as opposed to every ten minutes or so for the chaffinch.

Perfect whistle

Despite their apparently limited vocal repertoire, bullfinches are perfectly capable of learning to imitate sounds. In Germany, there was for many years a tradition for teaching captive-bred bullfinches to whistle various folk-tunes, with keenly fought contests to see which was the most perfect performance. Some recordings of these bullfinches survive today and have to be heard to be believed.

Mutual friends

The majority of fledgling birds, at least of small species, have little or nothing to do with their siblings when they have left the parental territory. In bullfinches, however, something astonishing happens. Once about 6–7 weeks old, each bird forms an intimate association with one of its siblings, male or female. The birds caress each other's beaks, feed one another and even invite each other to mate (although no mating takes place). Next spring they form normal associations with the opposite sex.

Mutual displays

Male and female bullfinches have a very intimate and exclusive relationship. The male's song, which is very quiet and highly individual, has no territorial function but is used instead to beguile the female early in the breeding season.

Among the other displays, both sexes play an equal part. When the birds meet for the first time, they indulge in 'bill-caressing'. The male approaches with its tail turned towards the female, puffs out its breast feathers and touches the female's bill with its own, then quickly turns away; if the female is interested, it will do the same. This display quickly becomes courtship feeding (feeding of the female by the male), once the pair is established.

A hidden secret

Bullfinches possess a particularly unusual physical feature: the male's sperm is unlike that of any other small, relatively advanced bird in the world. The details of the differences are not relevant here, but the learned conclusion is that there is a great deal less 'sperm competition' – competition for access to willing females – in this monogamous bird than is usual among most of its peers, and that its sperm structure has evolved as a result.

Ashes to ashes

Ash seeds are one of the bullfinch's key foods in the non-breeding season. Studies on bullfinches eating the buds of fruit trees in orchards, where they can become a pest, have shown that raids on orchards happen far less often when the ash trees have had a productive seeding season.

▽ *While most finches are strictly seed-eaters, bullfinches take a great deal of soft plant matter, including buds, flowers and berries.*

Hawfinch

Species: *Coccothraustes coccothraustes*
Family: Fringillidae

IDENTIFICATION This big, burly songbird (18 cm/7 in, much larger than a greenfinch) has a thick neck and an enormous bill, which – combined with the short tail – makes it look top-heavy. Look out for the black throat and face mask, grey neck, chocolate-brown back and pinkish breast, rump and tail. It shows prominent, broad, pale wing-bars in flight, particularly from below, and a white-tipped tail. At rest, the wing-bars appear like a large blob of white/pale brown.

The *male* is larger and brighter-coloured than the female. The main difference is in the flight feathers, with glossy blue-black secondaries twisted at the tip. Its bill is blue-grey in summer, yellowish in winter.

The *female* is less black on the face than the male, and its plumage is somewhat duller. Its secondary feathers make a grey wing-panel. Its bill changes, like the male's.

The *juvenile* (early to late summer) is an odd-looking finch with an adult's shape, but it is dark brown, with large spots on the breast.

△ *Male and female hawfinches look much alike, but note the grey panel on the wing that distinguishes this bird as a female.*

SHAPE AND CHARACTER If seen clearly, the hawfinch is unmistakable, with its outsize bill, hefty body and short tail. It flies with heavy undulations and super-fast wing-beats that seem only just to keep it airborne. It has a tendency to fly high up. A notably shy bird, it takes off at the slightest hint of danger and can be a difficult bird to see. It feeds unobtrusively on the ground, as well as in tree crowns. It is sparingly sociable, feeding in small flocks, although larger concentrations occur at roost.

VOICE It is a quiet species, but its calls are quite distinctive. Most common is an explosive (not necessarily loud) 'pix', heard in flight or when settled. This resembles the robin's tick, but is not as clean and has a slightly spitting quality. The hawfinch has a pathetic, strained, jerky song, delivered from the treetops, but still barely audible; it includes various 'swee' and 'swit' notes.

HABITAT This is primarily a bird of broad-leaved woodland (beech, oak, hornbeam) that visits fruit trees and feeders in orchard, gardens and parks.

FOOD It specializes on large, hard-shelled seeds, such as hornbeam, cherry, maple, beech and (in southern Europe) olives. These it takes in season, from summer to spring, but also eats buds, shoots and, surprisingly, significant numbers of caterpillars and other insects in the breeding season.

IN THE GARDEN It visits bird tables for hard-shelled nuts, seeds and fruit. It is shy and easily disturbed.

BREEDING It tends to select mature trees and has quite specific requirements. The normal clutch is four to five eggs, incubated mainly by the female for 11–13 days. The young leave the nest at about 14 days old and, in contrast to most other finches, constitute the adults' single breeding attempt in a season.

MIGRATION This bird is not a great migrant, with most populations staying put in an area. Some individuals, especially northern birds (juvenile and females) travel south-west in the autumn, but don't go far.

ABUNDANCE The hawfinch is fairly common in continental Europe, but is a rare garden visitor in Britain.

HAWFINCH CONFIDENTIAL

Big bill

The hawfinch's bill looks like – and is – a formidable tool. It enables the bird to specialize in eating very hard seeds that are quite unmanageable for finches with smaller bills. Part of is strength is in the size, but much also depends on its internal structure. Both mandibles, upper and lower, are fitted with two horny, serrated knobs on their surface. The sets overlie each other on the midline of the bill. Seeds are held between these four knobs so that the suture runs from front to back and the powerful jaw muscles close the bill. Enormous pressure is concentrated at the four knobs, localizing the force and making a rupture.

The force exerted by a hawfinch's bill is truly astonishing: it has been measured at more than 50 kg (110 lb) – that is 900 times the bird's own body weight.

Socialites and loners

For some reason, hawfinches have two systems of breeding. One is colonial, breeding in loose groups of up to seven nests, the nearest just 3.7 m (12 ft) apart; such birds only defend the immediate vicinity of the nest and forage in groups. The other system is fiercely solitary, the pair defending an area of about 2,000 sq m (2,400 sq yards) around the nest, foraging alone and not allowing any other hawfinch near. Nobody knows the reason behind this well-defined dichotomy.

▽ *The enormous bill is the hawfinch's defining characteristic, enabling it to crack nuts that are too hard for other birds to tackle.*

Rose-ringed parakeet

Species: ***Psittacula krameri***
Family: Psittacidae

△ *Rose-ringed parakeets depend on feeding stations for their survival, at least in the northernmost parts of their range.*

IDENTIFICATION This green parrot (38–42 cm/ 15–16½ in) is the size of a large pigeon, but slimmer. It has a long, pointed tail, narrow pointed wings and fast flight. The bill, red in adults, is large and typically curved in parrot style, with a large upper mandible. It is mainly green, with a blue tinge on the tail and darker main flight feathers.

The *male* has a black throat, the colour of which extends across the neck, narrowing and becoming bordered with a deep rose-pink. The nape is powder-blue.

The *female* lacks the above adornments, but might have the hint of a neck collar.

The *juvenile* (early spring to early autumn) is similar to the female, but with a pale-brown bill and a shorter tail.

SHAPE AND CHARACTER The rose-ringed parakeet's long-tailed parrot shape is unmistakable. It flies with distinctive, shallow wing-beats, but is fast and exceedingly manoeuvrable; it flies straight and doesn't undulate. It is sociable and often very noisy. With its green plumage, it easily disappears into the treetops where it does most of its feeding.

VOICE It utters a loud, piercing screech – 'ke-ack-ke-ack' – which may be repeated. It also makes various quieter conversational sounds.

HABITAT In Europe it is associated with towns, cities, parks and farmland. Roosts are usually situated in isolated lines of tall trees.

FOOD The rose-ringed parakeet eats all sorts of plant food, including seeds, fruits, leaves, flowers, berries, buds and nectar. It is perfectly capable of feeding on grain and has the potential to become a serious pest. It is very agile among tree branches.

IN THE GARDEN This parakeet will visit feeding stations (hanging or flat) without invitation to feed on nuts, seeds and even garden scraps. It will also use nest-boxes, but this should probably be discouraged.

BREEDING Despite its tropical provenance, it apparently breeds at almost any time between mid-winter and early summer, at least in Britain. Its nest site is a hole in a tree. It lays three to four white eggs, which are incubated by the female, fed by the male, for 22–24 days. The young leave the nest at 40–50 days and are dependent for another month or so.

MIGRATION This bird is not migratory.

ABUNDANCE It is very patchily distributed, but common where it occurs.

ROSE-RINGED PARAKEET

CONFIDENTIAL

What's it doing here?

The rose-ringed parakeet is not native to Europe. Its native range is the Sahel sub-Saharan belt of Africa, and similar latitudes in Asia, particularly India. It has the distinction of being the only parrot in the world to be found naturally on both continents. Its appearance in Europe is artificial; it was introduced to several urban areas in Britain, Belgium, the Netherlands and Germany between the early 1960s and 1980s.

So how does it manage here, especially considering that the British populations are the most northerly of any parrot in the world? The answer is simple: bird feeding stations. In mild winter weather this adaptable species can easily thrive, but in harsh conditions it is almost certainly due to the kind provision of feeding stations that this unexpected interloper has the chance. Remarkably, it seems to have a breeding season more in keeping with conditions in its native range, often nesting in the winter months of January and February.

Riotous roosts

In common with starlings, rose-ringed parakeets roost communally, gathering in trees in the evening and often commuting from some distance away. Prior to settling down, the collective screeching of these birds can be a remarkable noise.

Where the rose-ringed parakeet is abundant, as it is in India, the roost can be enormous: more than 100,000 birds have been estimated at a single roost. If the parakeet continues to increase in Europe, as it has been doing relentlessly, that could be a sound of the future...

▽ *The rose-ringed parakeet is pretty unmistakable, and it still feels incongruous to see it flying wild in Europe.*

Sparrowhawk

Species: *Accipiter nisus*
Family: Accipitridae

IDENTIFICATION The sparrowhawk (30–41 cm/ 12–16 in, about the size of a pigeon) is usually the most common predatory bird in gardens. It is dark above, pale below, with dense barring across the underside. The yellow talons and bill base are obvious. Its eyes have a wild, demented look. It has a pale supercilium (eyebrow) and often a spot on the back of its neck, plus a wingspan of 60–80 cm (24–31 in).

The *male* is smaller than the female. The upper parts are dark grey, the bars on the underside orange-brown, and the cheek orange-brown.

The *female* is dark grey-brown above; its underparts are white, with dark bars.

The *juvenile* (mid-summer to mid-autumn) is reddish-brown above, with coarse, haphazard barring on its chest.

SHAPE AND CHARACTER The combination of long tail and rather short, blunt wings distinguishes this from most other birds of prey except the goshawk. It is usually seen in flight, soaring upwards, when it alternates glides with bursts of two to three flaps; it is often accompanied by mobbing starling or pigeon flocks. It is also seen as it bursts upon bird feeders at high speed. You can often tell that a sparrowhawk is about before the bird appears: birds give high-pitched alarm calls and dive as one towards the ground.

VOICE It utters a hoarse, complaining 'kek-kek-kek', mainly heard from birds near the nest.

HABITAT The sparrowhawk is attracted to the garden scene by bird feeders. It breeds on the edge of woodland, but hunts along clearings and hedges. It spends much time concealed in the crowns of trees, watching for prey.

FOOD It eats live birds that range in size from tits to pigeons, and which are caught by its stealthy approach and rapid strike; victims are snatched with the feet.

BREEDING It begins breeding in late spring, and there is one clutch of four to five eggs. The female incubates the eggs for 33 days, while the hard-working male brings all its food. When the eggs hatch, the male provides for all the chicks as well. Only when they are well grown do both parents hunt. The young leave the nest after 27–31 days, but are dependent for a long time after this.

MIGRATION This bird is resident over much of Europe, but individuals may wander. Sparrowhawks in Scandinavia are summer visitors.

ABUNDANCE It is a common and widespread bird.

◁ *A sparrowhawk soars on its fairly broad wings. Note the long tail.*

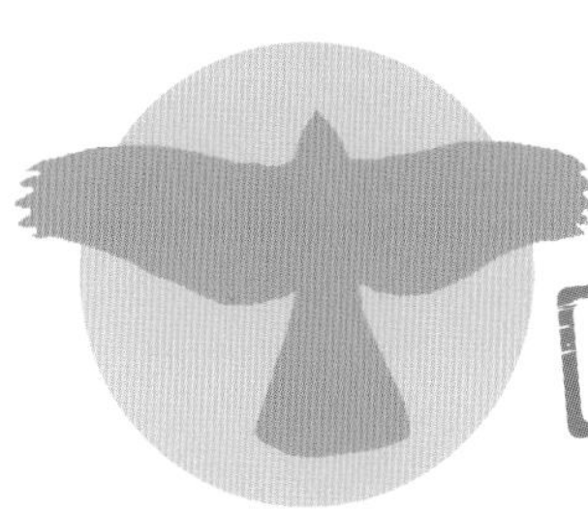

SPARROWHAWK

CONFIDENTIAL

Screens of all kinds

Hunting success for a sparrowhawk depends on taking birds – a highly mobile, manoeuvrable prey – by surprise. This isn't easy. As a flying predator, the sparrowhawk needs to accelerate to considerable speed before it actually reaches potential prey and has a chance to grab it. This it does by approaching a victim's regular feeding place – a bird feeder or seed-filled field – concealed behind some kind of barrier. In the countryside, the hunter often approaches along the opposite side of a hedgerow, before flipping over and making its final dash. In gardens the ideal screen is a wall, hedge or fence, but birds also use washing lines, cars, even moving people – anything to get a slight advantage over their potential prey.

Male and female size

A female sparrowhawk can be as much as 25 per cent bigger than her breeding mate. This bald fact masks an astonishing reality: the size difference is among the greatest for any bird in the world and ensures that, away from the nest, the sexes segregate. For example, females tend to live in open country and subsist on larger prey such as starlings, pigeons and thrushes, while males are more prevalent in woodland, where they specialize on tits and finches.

Why the difference? This has proven almost impossible to explain. The male's small size makes it a particularly agile hunter in the breeding season and enables it to provide for the family. But why should the female be so big? Its bulk might help it protect the nest, but nobody really knows.

Late breeding

On average sparrowhawks lay their single clutch of eggs in late spring, much later than most resident birds – including, significantly, those species on which they prey. Not surprisingly, this timing is deliberate. By the time the sparrowhawk chicks hatch, in early summer, gardens and woodlands are full of small birds (especially tits) that have just fledged. These fledglings are inexperienced and slow, and thus much easier to catch than the adults, making them perfect fodder for the sparrowhawk to bring to its own growing chicks.

▽ A sparrowhawk feeds its nestlings. When the chicks are young, the adult usually dismembers food for them.

△ On the whole, and especially in winter, male and female sparrowhawks hunt slightly different prey. It is males that take small woodland birds such as tits.

Pheasant

Species: *Phasianus colchicus*
Family: Phasianidae

IDENTIFICATION This is a distinctive game bird (53–89 cm/20¾–35 in) that lives mainly on the ground.

The *male* is spectacular and unmistakable, with a red face-wattle, bottle-green neck, glorious coppery tint to its dark-spotted plumage and an opulent long tail with bold striping.

The *female* is smaller and more soberly coloured, a mixture of speckled browns, still with a distinctive long tail.

The *juvenile* (early to mid-summer) is small with a short tail.

SHAPE AND CHARACTER The pheasant's size and shape are unmistakable. It is easy to see in fields, and often runs away from cars on rural roads. It flies off noisily from cover with a burst of its broad wings, its tail trailing behind, then alternates a flapping and gliding flight. It is often noisy at night as it roosts in the low branches of trees. It is sociable, living in small groups, often of a single sex.

VOICE The pheasant utters a familiar coughing crow: 'kok-koch'. In display the male flaps its wings loudly at the same time as it crows.

HABITAT This bird thrives in farmland areas with a combination of open fields and patches of woodland.

FOOD It eats mainly seeds and grain, but also berries, leaves and other plant matter. It takes small insects, especially ants and beetles, in the breeding season. It obtains food on the ground, often by scratching the litter with its feet; it will also dig with its bill to a depth of 8 cm (3 in).

IN THE GARDEN It will come to grain, usually feeding on the ground, but sometimes flies up to open bird tables.

BREEDING The pheasant's well-hidden nest is usually placed on the ground within the cover of tall plants. The clutch contains the high number of 10–14 eggs, which are brown and well concealed. Usually only the female incubates for 23–28 days and looks after the young. The hatched young leave the nest straight away and can feed themselves; they can fly after another 10–14 days.

MIGRATION This is a sedentary bird.

ABUNDANCE The pheasant is common and widespread.

◁ *In contrast to the spectacular male, the female pheasant is a modestly coloured bird – a necessity for camouflage when at the nest.*

▷ *Despite all the gaudy colours, what a female pheasant looks for is the red wattles on a male's face. Their size is an indication of fighting ability.*

Male harem

Pheasants don't always operate the standard breeding arrangement of birds. Instead the male, if it can, acquires a small harem of females (two to three), whose members build nests within or close to its territory. The successful male mates with each female, but does not contribute to looking after the young. Instead, it is taken up with defending its territory from other males.

The pheasant's system is a meritocracy and skirmishes between males over territory are common. The best males acquire a harem, while lower-quality birds may not gain a territory and breed at all; interestingly, the dominance of a bird is directly correlated to the length of the spurs on its legs. The losing birds sometimes stomp vengefully around the edge of a successful male's territory and even disrupt the nesting attempts of one of the females. However, most males probably accept a single female as a mate and have a monogamous breeding season; in which case the male will regularly help to look after the chicks when they have hatched.

Red-faced

You might think that the male pheasant's gaudy plumage and impressive tail should be more than enough to impress a visiting female and entice it into the harem. But it isn't. What females are looking for is a lot more subtle; they look at the face.

Two features stand out. First, the size of the red face-wattle is important. Experiments have shown that males with larger wattles have a greater fighting ability and are thus able to keep hold of their territory throughout the season. Second, the small black spots within the wattle are significant. When a male is excited and inflates its red wattle, the black spots become more obvious. Apparently, the width of the spots bears a direct relationship to the physical condition of the bird, and specifically its testosterone levels. So much information in the face – it's enough to make you blush.

Roosting

Although it is a ground-living bird, the pheasant prefers to take to the trees at night – it's a lot safer. Most individuals find a suitable tree on the edge of a wood, and a number of birds will often roost together in the same tree. In very cold conditions, they might huddle together to keep warm, but normally they keep their distance.

You might expect them to go to roost with a degree of secrecy, slipping into the branches without giving away their whereabouts. In fact, they make a bit of a song and dance of it. The males, especially, routinely give off loud crowing calls into the darkness. They will even call as they flutter, stepwise, into the higher branches of their chosen tree. In fact, on moonlit nights they give loud, echoing calls apparently at will – it's a wonder that any of them get any sleep.

◁ Some male pheasants have a white ring around the neck, indicative of an origin in China rather than the Near East.

▷ When roosting, pheasants take to the trees for safety, but they still look incongruous above ground.

Herring gull

Species: *Larus argentatus*
Family: Laridae

IDENTIFICATION This is a large, noisy, boorish gull (55–67 cm/21½–26½ in) of seaside towns in northern Europe. It has pink legs and a wingspan of 1.4 m (4½ ft).

The *adult breeding* (late winter to early autumn) has a pale-grey back, with white head, neck and underparts. Its bill is yellow with an orange spot; it has a pale eye and a somewhat 'angry', frowning expression. Its wings are grey with black tips and several white spots (mirrors).

The *adult non-breeding* (mid-autumn to early spring) is similar, but the head is speckled with brown spots.

The *immature* (years 1–3) begins as spotted brown all over and becomes progressively more like the adult as it grows. The range of plumages is confusing.

SHAPE AND CHARACTER The herring gull is large, with a heavy bill and broad wings; it is not as sleek as many other gulls. It flies expertly, and can manoeuvre to snatch scraps off a street, but also soars and glides frequently, wheeling up to great heights like a bird of prey. It is tame and at ease with the activities of people.

VOICE Its most familiar noises are various wails and grunts, a typical seaside background. It gives a loud series of wails at laughing pace, the familiar 'long call'. It also makes a slow throaty 'ga-ga-ga' with a slightly worried tone.

HABITAT This is primarily a bird of sea coasts, an abundant garden resident in seaside towns and cities.

FOOD The herring gull is omnivorous (including fish and shellfish from the sea). In towns it is a scavenger of mainly animal products, including offal and road-kills. It is frequently seen at rubbish dumps, and in gardens will take scraps and bread.

△ *Here comes trouble. The herring gull can become disturbingly bold in some coastal towns.*

IN THE GARDEN The herring gull needs no invitation to take scraps from lawns, feeding stations or rubbish in seaside towns.

BREEDING It lays a single clutch of three eggs. It usually breeds in colonies of varying size, each pair holding a small territory within the colony structure. Its nest is a large mound of vegetation generally placed on the ground, although in recent years many herring gulls have taken to nesting on flat roofs. The young hatch after 28–30 days and leave the nest 35–40 days later.

MIGRATION The herring gull is resident, although many Scandinavian birds move to the British Isles for the winter.

ABUNDANCE The herring gull is very common on coasts; much less so inland.

HERRING GULL

CONFIDENTIAL

Female–female pairs

Unusually among birds, female herring gulls in a colony quite regularly form pair-bonds with the same sex, making a nest and attempting to bring up chicks. One or both birds copulate with nearby males, but incubation and all chick-care are divided between the two females. Interestingly, chicks from the nests of female–female pairs rarely survive, regardless of the quality of their genetic material.

Orange communications

The orange spot on a herring gull's yellow bill is there for a purpose: it acts to stimulate the chicks to beg the adults for food. Once the chick pecks the bill, the adult is in turn stimulated to regurgitate half-digested food for its youngster to feed upon. The simple function of the orange spot, revealed by Niko Tinbergen in the 1950s, was one of the first discoveries in the fledgling science of the study of animal behaviour.

But it isn't only chicks that peck at a herring gull's orange spot. During adult courtship, the female begs in exactly the same way – and with the same result. The male regurgitates food and the female eats the offering; feeding by the male makes an important contribution to getting the female into condition for egg-laying.

Swapping families

Incredible as it may seem, some young herring gulls abandon their own parents in favour of a family next door. Such a step is potentially hazardous, since gull chicks straying into the 'wrong' territory are routinely attacked and killed by adults. Nevertheless, it seems the chicks take this extreme step out of necessity, if their parents are failing to provide them with enough food to survive. Starving, they seek out an alternative territory in which the chicks are slightly younger, and just hope for the best.

Aggression towards humans

Herring gulls in towns have little fear of humans and treat us without much deference. In some towns in England, breeding gulls have taken to attacking people who intrude on their territory. In one English town, the local gulls habitually attack postal workers trying to deliver mail – but apparently only men, not women.

Dropping stones

Herring gulls are adaptable and intelligent birds, as anyone who has seen their ingenious method of breaking open seashells on beaches would testify. They take the prey in the bill, fly to a moderate height and then drop the shellfish onto a hard surface to break it open. Mind you, they frequently drop the hard shells on sand, leaving them intact, and may carry on this fruitless procedure for minutes on end.

▷ *Young herring gulls start off a mottled brown colour, like this bird a few months old, and attain adult plumage only in their fourth year.*

Index

G

H

I

Acknowledgements

The publishers would like to thank David Tipling for his kind help in obtaining photographs for publication in this book.

Alamy A Esparraga 170; Alex Fieldhouse 108, 154; Andrew Darrington 17, 64, 97, 146; Antje Schulte - Birds 150; Arco Images GmbH 31 b, 82, 99, 144; Arterra Picture Library 2, 33, 101, 158; Bengt Lundberg 34; Birdpix 138, 168, 186; blickwinkel 18, 27, 69, 92, 106, 117, 130, 140, 141, 179; Brian McGeough 152; British Garden Wildlife 85; Buschkind 172; Chris Gomersall 54, 126; Dave and Sigrun Tollerton 94; Dave Bevan 95; David Chapman 1, 37, 81, 136, 151, 163, 176; David Kilbey 145; Derek Croucher 183; Felipe Rodriguez 8; Frank Blackburn 20; Gary Beers 19; George Reszeter 171; Gordon Langsbury 90; Iain Davidson Photographic 15; imagebroker/Hans Lang 24; imagebroker/Ottfried Schreiter 84; INSADCO Photography 87; Jam World Images 116; Julie Thompson 55; Kevin Sawford 134; Lisa Moore 63; Michael Stubblefield 143; Pat Bennett 182; Peter Grimmett 155; Philip Mugridge 93 a; Rolf Nussbaumer Photography 113; tbkmedia.de 115; Terry Whittaker 125; The National Trust Photolibrary /Ian West 65; The Photolibrary Wales 159; Thomas Hanahoe 187; Tierfotoagentur 122; Tim Graham 13; Vasiliy Vishnevskiy 127, 175; Vaughan Ryall 60; Wildlife GmbH 2, 107; Willie Sator 14; WoodyStock 75. **Chris Gomersall** 135. **David Kjaer** 21, 32, 74, 91, 123, 132, 161, 177, 181. **David Tipling** 4, 6, 9, 16, 23, 28, 31 a, 35, 38, 40, 42, 43, 44, 46, 51, 53, 57, 61, 67, 70, 76, 77, 80, 89, 102, 105, 109, 114, 118, 119, 131, 147, 157, 162, 166, 173, 184. **FLPA** Imagebroker 103; Neil Bowman 47; Phil McLean 160. **Fotolia** Andrew Bruton 12; avdwolde 128; Christian Beudez 121; Colette 178; Filev 104; Grégorie Landru 59; John Barber 72; John Saunders 96; Marcin Perkowski 112, 148, 169; Megan Lorenz 22; Michael Ushakov 56; Ornitolog82 39; Rick Thornton 50; robag 100; Samuele G 165; Sharpshot 142; Stephen Jones 58; vchphoto 88. **John Robinson** 48. **Nature Photographers** Paul Sterry 62. **Nature Picture Library** Andy Sands 41; Dave Bevan 68; David Kjaer 86; Kim Taylor 25; Laurent Geslin 11; Markus Varesvuo 139, 185; Mike Wilkes 71, 73; Roger Powell 26; Steve Knell 52; William Osborn 78, 79. **Photolibrary Group** Fotosearch 66; Imagebroker/ Jörn Friederich 129; Juniors Bildarchiv 137; Nature Picture Library 7 r, 133; OSF/David Boag 7 l, 98; OSF/David Courtenay 49, 153; OSF/Roland Mayr 36; Picture Press/Rolf Mueller 149; Picture Press/Willi Rolfes 110. **Roger Tidman** 45, 180, 83. **RSPB Images** David Kjaer 93 b; Richard Brooks 174. **Still Pictures** blickwinkel 120, 124. **SuperStock** age fotostock 111, 164, 167; Mauritius 156. **Warren Photographic** 10.

Executive Editor – **Trevor Davies**
Editor – **Ruth Wiseall**
Executive Art Editor – **Penny Stock**
Designer - **Ginny Zeal**
Picture Research Manager – **Giulia Hetherington**
Production Assistant – **Vera Janke**